Wildflowers
of the Central South

Wildflowers of the Central South

by

Thomas E. Hemmerly

VANDERBILT UNIVERSITY PRESS
Nashville, Tennessee
1990

All photographs are by the author except for the following, which are identified either by page number and order (A,B,C) on that page, or by figure: Carol C. Baskin (106B), Sharon K. Bracy (Fig. 4), Bertha Chrietzberg (94C, 96A), Milo Guthrie (22B, 36D, 38A, 48C, 56B, 66C), Landon McKinney (72A), Melville R. McMillan (32C), Elsie Quarterman (20C, 22C), Jack Ross (author, back cover), Tennessee Department of Conservation (Fig. 5), Paul Somers (36A), and Sandy Suddarth (70C). Susan J. Harrison executed the line drawings and Tom Ventress the map of the Central South.

Library of Congress Cataloging-in-Publication Data

Hemmerly, Thomas E. (Thomas Ellsworth), 1932–
Wildflowers of the Central South / by Thomas E. Hemmerly.
p. cm.
Includes bibliographical references and index.
Summary: Briefly examines the ecology of the central South, defined as a small area of northern Alabama, central Tennessee, and a portion of south-central Kentucky, and describes over 250 plant species found in that area.
ISBN 0-8265-1240-2 (paper)
1. Wild flowers--Tennessee--Identification. 2. Wild flowers--Alabama--Identification 3. Wild flowers--Kentucky--Identification. 4. Wild flowers--Tennessee--Pictorial works. 5. Wild flowers--Alabama--Pictorial works. 6. Wild flowers--Kentucky--Pictorial works. [1. Wild flowers--Southern States.]
I. Title. II. Title: Wild flowers of the Central South.
III. Title: Central South.
QK187.H46 1990
582.13'0976--dc20

90-38197
CIP
AC

Published in 1990 by the Vanderbilt University Press
Printed in the United States of America

In memory of
my mother
Jessie Mai Cashdollar Hemmerly
1905 – 1976

Contents

Preface ix

Introduction 1

Part 1 — Ecology of the Central South 3

Part 2 — Ecology of Cedar Glades 7

Part 3 — Identifying Wildflowers 13

Wildflowers Illustrated 19

Part 1 — Monocots 20

Part 2 — Dicots 41

Bibliography 115

Index 117

Preface

Why another wildflower book? Even though many attractive and helpful wildflower guides are available, most are so broad in geographical coverage that many local plants are not included. This is especially true for wildflowers of the cedar glades, important ecosystems found principally in the Central South. These plants, many of which do not occur elsewhere, are given special attention here.

Conservation should result from expanded information and awareness of any aspect of natural history. After all, does knowledge not lead to appreciation, and appreciation to the desire to protect? As the noted ecologist Paul Sears observed, "People seldom try to save anything that they haven't yet learned to love." Many rare and endemic plant species (and their habitats) of our area are threatened and require care and protection to ensure their survival.

This book is intended for the serious student of nature, regardless of background. Technical terms are kept to a minimum and are defined as necessary. However, a determined effort has been made to assure that the guide is scientifically accurate. In the bibliography, references are provided that will be useful for the average naturalist and the more advanced botanist as well.

Each of the more than 150 species of wildflowers illustrated is presented in the context of its environment and the region in which it occurs. Additional species similar or related to them are also described. Furthermore, information on medicinal, economic, and even some horticultural uses is included.

The author hopes that this small volume will contribute to an understanding and appreciation of the wildflowers of the Central South.

This book is the result of a grass roots effort to prepare a wildflower guide for local use. Toward that end, "The Committee," composed of Bertha Chrietzberg, Milo Guthrie, Landon McKinney, Ruth McMillan, and me, met periodically during 1985 and 1986 to screen and select transparencies as illustrations for such a book. The awarding of a Non-Instructional Assignment Grant by Middle Tennessee State University for the spring semester of 1989 provided me the released time necessary to begin the writing. I hereby express appreciation to members of "The Committee," who graciously allowed me to assume authorship, to the N.I.A. Committee, to Biology Department chairman Dr. George Murphy, and to MTSU president Dr. Sam Ingram.

Drs. Carol C. Baskin, Jerry Baskin, Elsie Quarterman, and Paul Somers critically reviewed substantial portions of the manuscript. Dr. Somers and Dr. Kurt Blum also helped in the identification of several wildflowers. I greatly appreciate their efforts to improve the book; however, I take full responsibility for any errors or deficiencies.

Mr. John Poindexter and the staff of the Vanderbilt University Press, including designer Gary Gore, business manager Jane Tinsley, copy editor Dimples Kellogg, and secretary Marge Cochrane, have all been enthusiastically involved in this publication. Its appearance and overall quality is a reflection of their respective skills.

Credit is due to Russell Blalock, Betsy Brown, Dawn M. Coffman, Ouida Plaisance, and Janice Stiles, all of whom were involved in typing the manuscript. Mary Gwyn Thompson also helped in its preparation.

Several persons gave me permission to study plants on their property – notably, Mr. Kenneth Evans and Mr. A. C. Bell, Jr., both of Murfreesboro. I am indebted to them and to numerous other individuals, largely unknown to me, across whose land I traipsed in the pursuit of photos.

Finally, I gratefully acknowledge the patience and encouragement of my wife, Beverly, during the completion of this project.

T.E.H.
April, 1990

Introduction

Illustrations

Map, Central South 2

Fig. 1 Mixed Forest, Radnor Lake, near Nashville, Tennessee 3

Fig. 2 Dissected Eastern Highland Rim, near Smithville, Tennessee 5

Fig. 3 Typical Cedar Glade, Cedars of Lebanon State Park, near Lebanon, Tennessee 8

Fig. 4 Cedar Glade in Winter, Cedars of Lebanon State Park, near Lebanon, Tennessee 8

Fig. 5 Dr. Elsie Quarterman, plant ecologist, Vanderbilt University 10

Fig. 6 Non-flowering Glade Plants: Witch's Butter, *Nostoc commune* (a cyanobacterium, left); Glade Moss, *Pleurochaete squarrosa* 10

Fig. 7 May Prairie, near Manchester, Tennessee 12

Fig. 8 Roadside Weeds: white Queen Anne's Lace, *Daucus carota*; Red Clover, *Trifolium pratense*; and, at right, Common Chicory, *Cichorium intybus* 13

Fig. 9 Leaf Structures and Variations 14

Fig. 10 Floral Structures and Arrangements 17

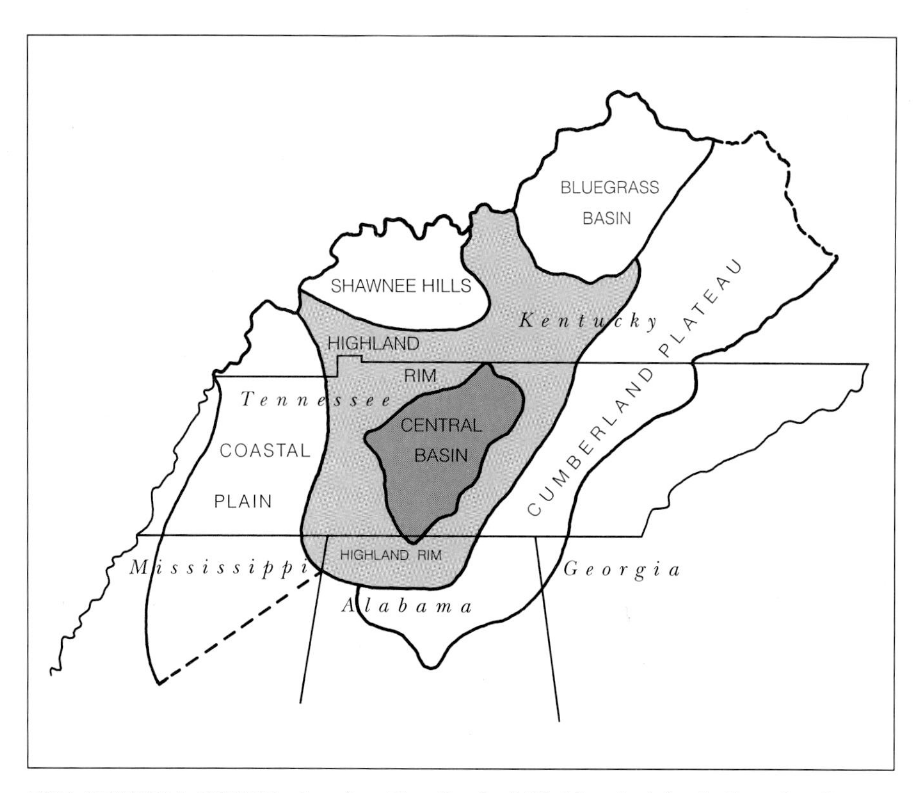

THE CENTRAL SOUTH showing the Central Highlands (shaded) and adjacent regions within which this book is most useful

Part 1
Ecology of the Central South

Living things, including the wildflowers featured in this book, do not exist in a vacuum. Plants act out the drama of biological phenomena, such as metabolism, growth, and flowering, on an environmental stage. Thus, to best appreciate the wildflowers of the Central South, we must consider how the ecological factors of the various regions interact to affect their distribution and their habitats.

Regions

At the heart of our area are the Central Highlands, which occupy all of middle Tennessee. From here they extend southward into northern Alabama and northward to include most of southcentral Kentucky. Designated by geographers a "physiographic province," the Highlands consist of two regions: a core area called the Central Basin and the surrounding Highland Rim.

The Central Basin is an elliptical depression some 120 miles long and 50 to 60 miles wide. With a mean (average) elevation of almost 600 feet, the Basin is generally flat or gently rolling except for "knobs," the hills decorating its outer edge. Examples of these knobs are the hills surrounding Radnor Lake (fig. 1), just south of Nashville.

The Highland Rim, a ring-shaped upland, encircles the Central Basin. It is relatively flat, except along its inner edge where it is deeply dissected by

Fig. 1 Mixed Forest, Radnor Lake

numerous streams. The view seen in figure 2 is typical of the interface between the Rim and the Basin below. The mean elevation of the Rim is approximately 1,200 feet.

The important geological features of much of the Highlands region are the extensive systems of underground drainage, consisting of caverns, sinkholes, and streams. The prevalence of soluble limestone is responsible for this karst topography.

Located peripherally to the Central Highlands are the other regions referred to in this volume, including the elevated Cumberland Plateau to the east and south, the low-lying Coastal Plain to the west, the Shawnee Hills to the northwest, and the Bluegrass Basin to the northeast. The Central Highlands, together with the Shawnee Hills and the Bluegrass Basin, comprise the Interior Low Plateau.

Climate

The Central South lies within a humid mesothermal climate region. Most years, there are four fairly distinct seasons. Summers are warm and humid; falls, mild and dry. Winters are wet, with temperatures alternating between frigid (sometimes below 0° F) and pleasantly mild. Springs are also quite rainy and typically consist of alternating warm and cool periods with rains decreasing as summer approaches. The frequent but unpredictable cool spells during this season are often referred to colloquially as "dogwood winter" (April), "blackberry winter" (May), or "cotton britches winter" (early June).

The mean annual temperature of the Central Highlands varies from 56° to 60° F. Within this region, the Central Basin is generally warmer than the Highland Rim. And the temperature on a typical day is several degrees cooler on the Cumberland Plateau (because of the higher elevation) and the areas north of the Highlands. However, it is not the mean temperatures but the minimum winter and maximum summer temperatures that are more significant in determining the distribution of plants. These, too, are affected by both altitude and latitude.

Despite considerable variation from year to year, precipitation for our area is usually between 45 and 55 inches per year. In general, rainfall decreases slightly toward the north and increases with elevation. As with temperature, moisture extremes (for example, droughts) are more important than means. Also, local conditions such as soil (type and depth) and slope of the land modify the actual amount of moisture available to plants.

Soil

Although the soils of the Central South are generally like those associated with deciduous forests throughout eastern North America, marked soil differences occur even within the Central Highlands. The soils of the Highland Rim, which develop largely from cherty limestone, are lighter, in both color and texture, and more acidic than the darker, heavier, more nearly neutral

Fig. 2 Dissected Eastern Highland Rim

limestone soils of the Central Basin. The cedar glades, which result from thin soils associated with this limestone, are the subject of Part 2.

The soils of the Cumberland Plateau tend to be rather acidic like those of the Rim, whereas the soils of the Bluegrass Region are similar to those of the Basin.

Vegetation

The Central South lies near the center of the once magnificent but now greatly modified temperate deciduous forest, which is the product of climatic and edaphic (soil) factors interacting over many centuries. The noted ecologist Dr. E. Lucy Braun, in her now classic *Deciduous Forests of Eastern North America*, recognized several regions of this forest. Of these, the Mixed Mesophytic Forest Region occupies the Cumberland Plateau. Probably the most representative of the deciduous forests, it is composed of forest communities dominated by a dozen or so tree species, including Sugar Maple, Sweet Buckeye, Basswood, Red Oak, White Oak, and Hemlock.

To the west of the Plateau, encompassing the remainder of our area, is the Western Mesophytic Forest Region. This is characterized by a variety of different climax forests determined by such local conditions as topography and special soil situations. On the more favorable (moist) sites are found mixed forests (fig. 1) much like those of the Plateau. On average or drier sites, various species of Oak (*Quercus*) and Hickory (*Carya*) dominate. In fact, A. W. Kuchler, in his map showing the potential vegetation of the U.S., includes practically all of our area west of the Plateau within the Oak-Hickory Region.

Plant Distribution

The distribution of wildflowers is determined by a myriad of biotic and abiotic factors. Ecologists often must devote considerable effort to untangling the web of interrelated influences responsible for the range of a given species.

Even considering the climatic variation already discussed, within our area soil pH (acid-alkaline balance) may be more important than climate in determining the distribution of some species. For example, many plants, such as members of the Orchid Family, require an acid soil for the growth of mycorrhizae, the mutualistic association formed between certain fungi and roots of plants. These wildflowers are predictably more likely to be found in the acid soils of the Rim or Plateau than in the Basin.

Part 2
Ecology of Cedar Glades

In the Central South, cedar glades occur primarily in the Central Basin of Tennessee where they are underlain by horizontal layers of thin-bedded Lebanon limestone. This substrate was formed some five hundred million years ago during the Ordovician geological period. The rock is now exposed or covered with rather shallow soil (fig. 3).

Cedar glades are open rocky, gravelly, or grassy areas; therefore, they are not synonymous with cedar forests. The designation "cedar" comes from the ever-present Eastern Red-cedar trees in the adjacent woods that surround and separate glades.

Our primary interest in cedar glades (or limestone glades) is the plant life occupying those unique habitats. However, before focusing on the botanical aspects, let us take a closer look at the physical environment of glades.

Geographical Extent

Murfreesboro, in Rutherford County, lies at the geographical center of Tennessee and also in the midst of the cedar glades of the Basin. In fact, the most extensive glades, and those with the largest number of endemic plant species, are in Rutherford County and two adjoining counties, Wilson (directly north) and Davidson (northwest). Unfortunately, they are populous, rapidly growing counties, and many glades have already been destroyed or are threatened by the encroachment of civilization. The Basin counties encircling the core glade region — Bedford, Cannon, Giles, Marshall, Maury, and Williamson — also contain scattered glades. Smaller glade areas, containing fewer cedar glade endemics, are found in eastern Tennessee, northern Alabama, Georgia, and southcentral Kentucky. West of the Mississippi River, cedar glades occur in the Ozarks of Missouri and Arkansas.

Microclimates and Adaptations

Open cedar glades include extreme microclimates that often contrast with the more moderate conditions of nearby non-glade areas. Winter temperatures in open glades are similar to those of adjacent forests. However, summer maximum temperatures at the soil surface often register 10° to 30° F above those measured in nearby protected habitats. Thin glade soils generally remain waterlogged during much of the wet winter (fig. 4) and early spring months, but rapid drying occurs during the warm months of the year, resulting in a virtual desert.

It is not surprising that glade plants, like many desert plants, have special means of coping with hot, dry summers. Many glade species, such as the Glade Stonecrop, flower early in spring and set seed before conditions become unfavorable. Others survive by having long roots that tap water from soil beneath

Fig. 3
Typical Cedar Glade, Cedars of Lebanon State Park (left)

Fig. 4
Cedar Glade in Winter, Cedars of Lebanon State Park (below)

the rock (Tennessee Coneflower) or by storing water in succulent leaves (Limestone Fame Flower).

Flora

Dr. Augustin Gattinger, a colorful German-born physician and botanist who settled in Nashville during the Civil War, was among the first to recognize and document the unique plant life of cedar glades. In his *Flora of Tennessee* (1901) he wrote of glades as "a natural conservatory that could fearlessly challenge any flower garden in the combined effect of gayety and luxuriance."

Since Gattinger, a succession of botanists has accepted his invitation: "For truth, my honored Tennessee friends, go and see, and learn to appreciate and to preserve such great ornaments of your native land." Foremost among those who have heeded his call is Dr. Elsie Quarterman, professor emerita of general biology, Vanderbilt University (fig. 5). Along with her graduate students and several other botanists, she has conducted investigations in cedar glades since the 1940s. Through the resulting scores of publications, the larger scientific community has been made aware of the importance of these ecosystems.

The prolific team of Dr. Carol Baskin and Dr. Jerry Baskin, both protégés of Dr. Quarterman and now at the University of Kentucky, is responsible for dozens of these papers. The Baskins's work is epitomized by their paper presented at the symposium "Vegetation and Flora of Tennessee" in March 1989. In it they brought together the current knowledge of the autecology (ecology of individual species) of the twenty-three known cedar glade endemics occurring in the Central Basin.

Plant Zonation

Cedar glade plants are subjected to various climatic and edaphic factors, all of which may influence their particular position within glades. The most vital factors include light intensity, temperature, and soil depth, which affects available moisture. Along a transect (line) from the center of an open glade (fig. 6) into the adjacent glade woods, a gradual transition in the magnitude of several environmental factors is apparent, as illustrated below:

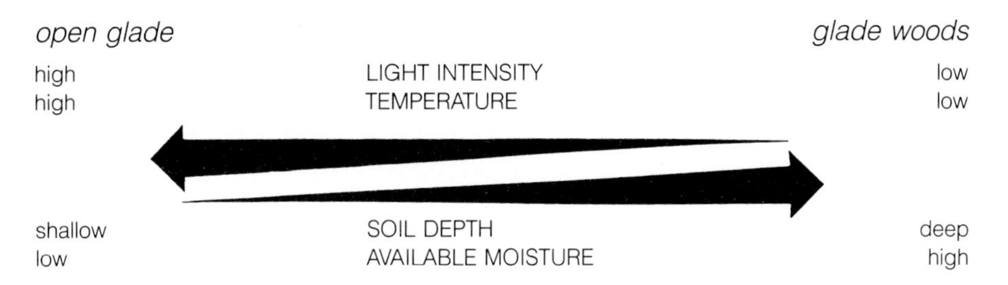

Over long periods of time, each species comes to occupy its particular position along this ecological spectrum. There its requirements for survival are met, making it possible to compete successfully with other plants.

Fig. 5
Dr. Elsie Quarterman (left)

Fig. 6
Non-flowering Glade Plants (below)

Following the lead of Dr. Quarterman, it is customary to recognize six plant zones, each based primarily on soil depth. These zones and some characteristic plants of each are shown in the chart below.

PLANT ZONATION IN CEDAR GLADES
(Numbers indicate page of illustration)

Zone/Soil Depth	*More Common Plant Species*	*Less Common Plant Species*
#1 EXPOSED ROCK	none	none
#2 GRAVELLY GLADES (0 to 2 in.)	Witch's Butter (10) Glade Sandwort (46) Nashville Glade-cress (56) Lime Stonecrop (58) Gattinger's Lobelia (98)	Limestone Fame Flower (44) Tennessee Milk-vetch (62) Violet Wood-sorrel (64) Price's Wood-sorrel (64) Glade Scorpion-weed (86)
#3 GRASSY GLADES (2 to 8 in.)	Glade Moss (10) Three-awn Grass (20) False Aloe (32) Blackberry-lily (34) Nashville Breadroot (60) Gattinger's Prairie-clover (62) St. John's-wort (68)	Star-grass (32) Eggleston's Violet (70) Glade Phlox (84) Rose Verbena (90) Glade Savory (90) Small's Ragwort (108) Tennessee Coneflower (112)
#4 SHRUBS (8 to 12 in.)	Aromatic Sumac (*Rhus aromatica*) Shrubby St. John's-wort (*Hypericum frondosum*) Coralberry (*Symphoricarpos orbiculatus*) Glade Privet (*Forestiera ligustrina*)	
#5 CEDAR WOODS (>12 in.)	Eastern Red-cedar (*Juniperus virginiana*) Hackberry (*Celtis laevigata*) Blue Ash (*Fraxinus quadrangulata*) Winged Elm (*Ulmus alata*)	
#6 OAK-HICKORY FOREST (>12 in.)	Oaks (*Quercus* species) Hickories (*Carya* species)	

Prairies

In addition to cedar glades, another type of natural opening occurs sporadically in our area. Called "barrens" by the English and "prairies" by the French, the grass-dominated openings stood in stark contrast to the extensive, almost continuous, forest found otherwise throughout most of the eastern third of the continent.

Because there is evidence that our small, scattered prairies were once continuous with the extensive prairies of the midcontinent, they are often referred to as "prairie relicts." As expected, these eastern prairies have many species in common with the midwestern prairies. In addition to grasses and sedges (collectively called graminoids), other flowering herbs, known as forbs, abound. Two plant families, Daisy (Asteraceae) and Pea (Fabaceae), are typically well represented among the forbs of prairies.

What is responsible for the persistence of these relict prairies? Varying explanations are necessary from site to site, but often local soil conditions are involved. In some cases, seasonal wetness seems to be a contributing factor.

Like cedar glades, eastern prairies are successional; that is, they are subject to being replaced by another type of biotic community. The eventual result of this process is, predictably, a deciduous forest.

Many prairies require disturbances, either natural or man-made, to preserve them. An example is May Prairie (fig. 7) near Manchester, on the eastern Highland Rim in middle Tennessee. Controlled periodic burning and brush cutting are practiced to prevent invasion by trees like Sweet Gum and Red Maple.

Fig. 7 May Prairie Near Manchester, Tennessee

Part 3
Identifying Wildflowers

A wildflower is any flowering plant (a member of the large group botanists call Angiosperms) that grows and reproduces without requiring cultivation. Plants that some would call weeds and also many woody plants (trees and shrubs) may be included under the umbrella term *wildflower.*

What is a weed? In general, weeds are invasive plants usually found in open, disturbed habitats. Naturally, crop fields are prime targets; weeds greatly damage agricultural crops. Most weeds are not native to North America, having been introduced from the Old World two or three hundred years ago (or earlier). Many weedy plants with their showy flowers add color and interest to our fields, forest edges, and roadsides, especially in summer and fall. Some attractive roadside weeds are seen in figure 8. Others appear throughout the book.

Since almost all woody plants native to our area are Angiosperms, they could logically be included in this book. As a matter of practicality, however, they have been excluded to allow more adequate coverage of the herbaceous (non-woody) wildflowers.

Identification of a wildflower may require careful consideration of the entire plant. All parts, including roots, stems, and leaves (vegetative parts) and flowers, fruits, and seeds (reproductive parts), are important and may be useful in its identification. Even so, leaf and flower features will be emphasized here.

Fig. 8 Roadside Weeds: Queen Anne's Lace, Red Clover, and Common Chicory

Leaves

Although they generally share a common function, namely, photosynthesis, leaves vary greatly in size, shape, and arrangement and in other ways that often reflect their environment. Some leaf variations (fig. 9) are listed here:

1. Arrangement on stem (9A)
   a. alternate–1 leaf per node
   b. opposite–2 leaves per node
   c. whorled–more than 2 leaves per node
2. Simple leaf (9B)
   a. blade–flat expanded portion
   b. petiole–leaf stalk (when absent, leaf is sessile)
   c. stipules–paired appendages at base of petiole (often absent)
3. Compound leaf (9C)–similar to simple leaf except blade is divided into leaflets (in some cases the leaflets are further divided)
4. Shape (9D)
5. Blade margins (9E)
6. Venation (9F)–the pattern of veins in a leaf

Fig. 9 Leaf Structures and Variations

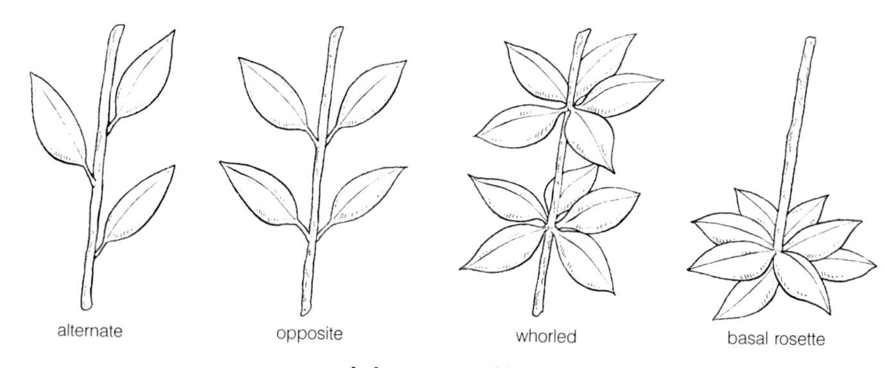

A. Arrangement of leaves

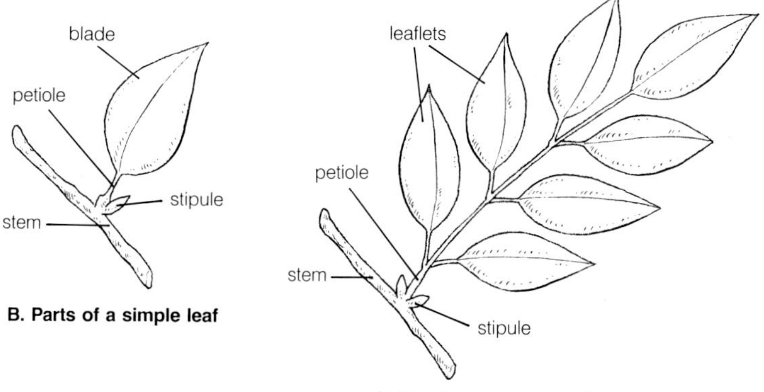

B. Parts of a simple leaf

C. Parts of a compound leaf

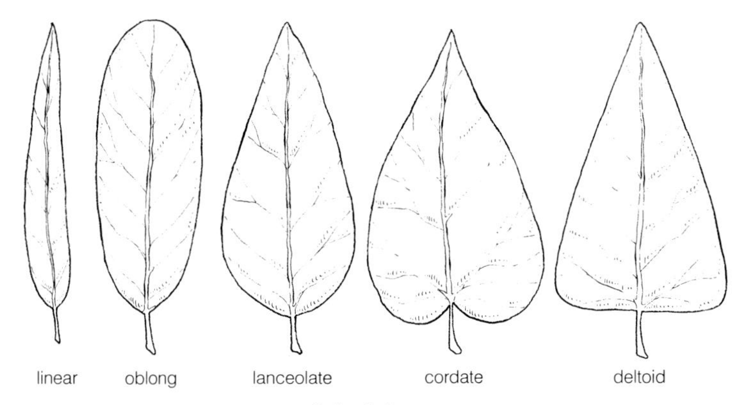

D. Leaf shapes

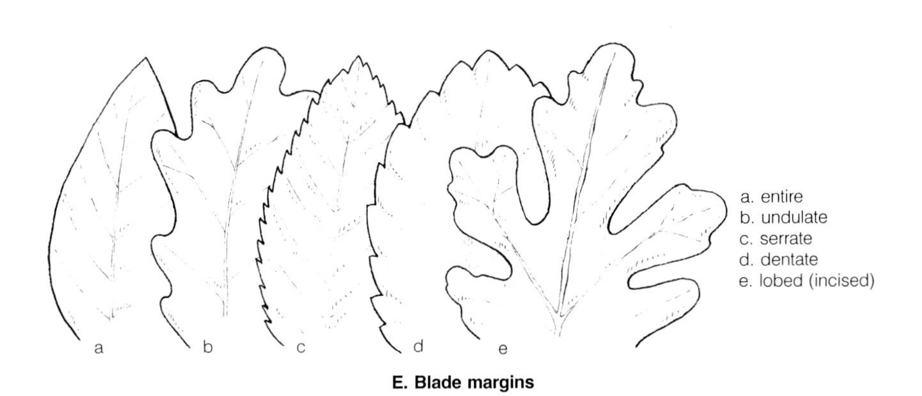

E. Blade margins

parallel veined

pinnately net-veined

palmately net-veined

F. Venation

Flowers

Angiosperms are characterized by the often beautiful structures we call flowers. More complex and specialized than a leaf, each flower is comparable to an entire branch of leaves.

A typical flower (if one exists) consists of four sets of parts, each set arranged in concentric rings called whorls (fig. 10A). Each whorl, defined by its relative position within the flower, has its own function to perform.

1. *Sepals* (collectively the calyx) are usually green and protect the bud; they are often fused to some degree.
2. *Petals* (collectively the corolla) are ordinarily the brightest part of the flower; they are often fused to form corolla tubes of various shapes.
3. *Stamens* are made up of the anther, which produces pollen grains, and the slender filament supporting the anther.
4. The *pistil* (usually one) is composed of the stigma (which receives pollen grains) and the ovary (which becomes a fruit) containing one or more ovules (each of which becomes a seed).

The great diversity in flower structure among the more than 250,000 species of Angiosperms is the result of deviations from and modifications of the typical flower. In some flowers, for example, stamens are present but pistils are absent; such a flower is said to be staminate in contrast to one that has only pistils, called a pistillate flower.

In some cases, flowers occur in particular arrangements known as inflorescences. Some of the more common ones are shown diagrammatically in figure 10B. In each case a stalk, the peduncle, supports a flower cluster. In some instances, one type of inflorescence is associated with a particular plant family. For example, the common daisylike "flowers" of the Composite Family (Asteraceae) are actually heads containing numerous tiny flowers.

Monocots vs. Dicots

All flowering plants belong to one or the other of two grand groups of Angiosperms: *Monocotyledones* (Monocots) or *Dicotyledones* (Dicots). These names are based on the number of cotyledons (one or two) possessed by the embryo of the seed. However, it is more practical to depend on leaf and/or flower features to decide if a plant is a Monocot or a Dicot.

The leaves of Monocots are typically parallel-veined (fig. 9F), whereas those of Dicots are either pinnately or palmately net-veined.

The flowers of Monocots are 3-merous, meaning that each or most whorls contain 3 parts (or twice 3). An example is a Trillium with 3 sepals, 3 petals, 6 stamens, and 1 pistil. The basic number for flowers of Dicots is something other than 3; usually they are either 4- or 5-merous.

Common vs. Scientific Names

The identification of a plant may involve determining one of its common names or its scientific name. Common names are essentially nicknames and

Fig. 10 Floral Structures and Arrangements

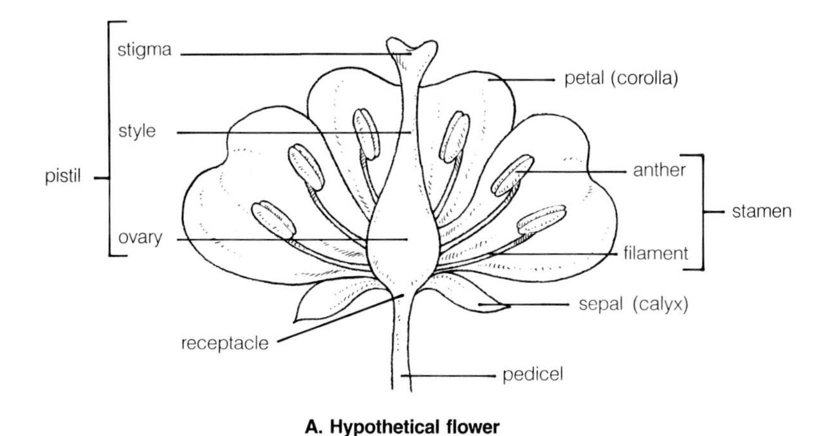

A. Hypothetical flower

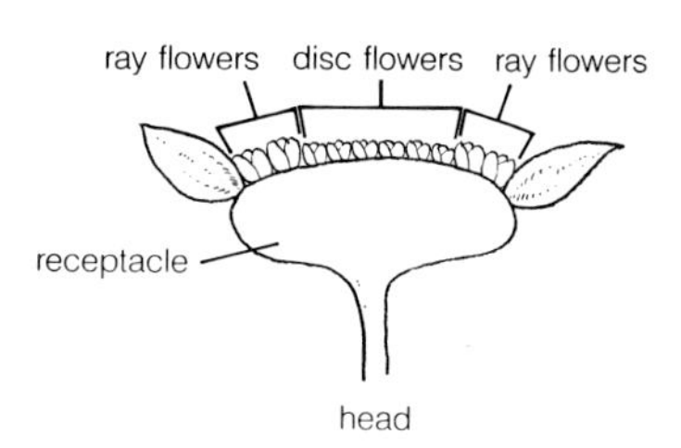

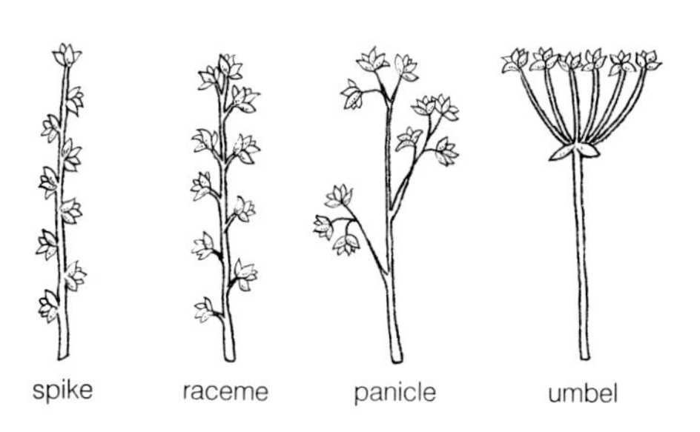

B. Inflorescences (diagrammatic)

as such are often fanciful or descriptive. Consider the shrub called Hearts-are-bursting-with-love, or Ladies'-tresses, a member of the Orchid Family. On the negative side, common names can be misleading. For example, Broom-sedge is actually a grass, not a sedge. And Reindeer Moss, found on the ground under cedar glade shrubs, is a lichen and, therefore, quite unlike any moss.

For these and other reasons, botanists use scientific names to avoid confusion and to be more exact. A scientific name is assigned when a new species is first discovered. Such a name must conform to the basic rules of taxonomy first established by Carolus Linnaeus, a Swedish botanist of the eighteenth century. In accordance with those rules, each scientific name is a binomial consisting of a genus name followed by a specific epithet (often called the species name). The first name is capitalized, the second is not, but both are italicized. The name (or initial) that follows indicates the person who first described the species and applied that name. An example is *Leavenworthia stylosa* Gray, named by the famous Harvard botanist Asa Gray.

In many cases, recognizable variation occurs within a species from one geographical region to another. Such a local group, if it varies markedly from other members of that species, may be formally designated as a variety or subspecies and named accordingly. This intraspecific name follows the genus name and specific epithet, resulting in a trinomial. Several examples are noted among the wildflowers of this book.

Format of This Book

The wildflowers featured in this volume are divided into the two great groups: the Monocots followed by the Dicots. Within each of them, families are arranged generally from the most primitive to the most specialized. Thus, the Monocot section begins with the Cattail Family, which includes the least specialized plants, and ends with the highly complex Orchid Family.

The information accompanying the photograph of each species of wildflower follows a reasonably consistent pattern. The common and scientific names are followed by the family names (the suffix "-aceae" denotes a plant family). An effort has been made to use the most recent scientific names. If a species is also widely known by an alternate name, it is shown in brackets. The first paragraph contains a brief description of the wildflower shown in the illustration. The second paragraph gives the estimated relative abundance, habitat, and geographic distribution. The third paragraph contains information on any similar species and often comments on economic or medicinal uses. Finally, the usual flowering time is given.

Wildflowers Illustrated

Part 1
Monocots

Common Cattail
Typha latifolia L.

Cattail Family
Typhaceae

Cattails, tall plants (up to about 6 ft.) with compact cylinders of tiny flowers, are quite unmistakable. The male flowers (already shed in this plate) pollinate the large mass of brownish female flowers below. The resulting seeds are airborne.

This plant is common in marshes and other wet places throughout the eastern U.S.

The rootstock is eaten by animals, including man. The pollen can be used as flour and the leaves to weave chair bottoms.

The Narrow-leaved Cattail, *T. angustifolia*, is smaller and less frequent but occupies similar habitats.

Flowering: April-June.

Broom-Sedge
Andropogon virginicus L.

Grass Family
Poaceae

Though its flowers are not showy, Broom-sedge, because of its distinctive golden-brown leaves and stems, is noticeable most of the year.

It grows throughout our area and beyond, especially in old fields like the one here, and frequently in cedar glades.

Several other species of *Andropogon,* especially Big Bluestem, *A. gerardi,* and similar grasses such as Little Bluestem, *Schizachyrium scoparium,* occur throughout our area, especially in limestone glades and barrens.

Flowering: August.

Long-Spiked Three-Awn Grass
Aristida longespica Poir.

Grass Family
Poaceae

The common name Three-awn Grass is applied to this large (40 species) genus of North American grasses. Note the 2 erect lateral awns only half (or less) the length of the central awn.

This species is widespread in poor soil throughout our area and beyond.

Although not taxonomically closely related, Poverty Grass, *Sporobolus vaginiflorus,* has a similar appearance when not in flower. The latter is much more common, but both are annual species in the grassy zone of cedar glades.

Flowering: September, October.

Common Cattail

Broom-Sedge

Long-Spiked Three-Awn Grass

Chufa *Cyperus esculenta* L.

Sedge Family Cyperaceae

Sedges are grasslike perennials typically found in moist places. The main stem when seen in cross section is triangular. In umbrella sedges (*Cyperus* species) the minute flowers, borne in scaly heads, occur in umbels. Chufa, 6 to 30 in. tall, has compact flower heads 1/4 to 1 in. long.

This weedy plant is found in ditches and on stream banks throughout our region.

Cyperus strigosus is also a common sedge in this area. In cedar glades, look for *C. inflexus,* which is only a few inches tall.

Chufa is also called Yellow Nut-grass because of its edible tubers. The Papyrus Plant, *C. papyrus,* was used by the ancient Egyptians to make paperlike writing material.

Flowering: August-October.

Frank's Sedge *Carex frankii* Kunth

Sedge Family Cyperaceae

Of the approximately 2,000 species of *Carex* worldwide, about 125 are in our area. The tiny staminate and pistillate flowers are typically borne in different parts of the same spike or in separate spikes altogether.

Frank's Sedge is one of nine species found in cedar glades. However, it occurs more often in moist calcareous woods and meadows throughout our area, especially in swamps and marshes of the Highland Rim.

Flowering: June-September.

Spike-Rush *Eleocharis compressa* Sulliv.

Sedge Family Cyperaceae

Of the more than 40 species of the genus in North America, only this one occurs in cedar glades. The specific epithet *compressa* refers to the flattened stems.

In addition to moist portions of cedar glades, it is also found in wetlands throughout the eastern two-thirds of North America.

Flowering: June, July.

Chufa

Frank's Sedge

Spike-Rush

Jack-in-the-Pulpit *Arisaema triphyllum* (L.) Schott.
Arum Family Araceae

This wildflower (12 to 18 in. tall) produces a spadix, a fleshy spike to which are attached tiny flowers. Surrounding the spadix is a modified bract (specialized leaf) called a spathe, which varies in color from green to striped maroon.

Look for this wildflower in deep moist soil in rich woods throughout the eastern U.S.

Green Dragon, *A. dracontium,* is a taller, more elongated plant with divided leaves and spadix extending several inches beyond the spathe.

The name Indian Turnip refers to the use of the corm (underground stem) for food. However, such use is not recommended because it contains needlelike crystals of oxalate that can produce painful irritation.

Flowering: March-May.

Slender Dayflower *Commelina erecta* L.
Spiderwort Family Commelinaceae

The blossoms of this upright perennial last only a day. Each flower has 2 prominent blue petals and a single transparent one.

This wildflower is common along roadsides and dry open sites throughout much of the eastern and central U.S.

The Asiatic Dayflower, *C. communis,* is a similar but weedy annual with stems tending to sprawl on the ground.

The stems and leaves of dayflowers may be eaten as potherbs.

Flowering: May-October.

Virginia Spiderwort *Tradescantia virginiana* L.
Spiderwort Family Commelinaceae

Spiderworts are perennials that produce terminal clusters of flowers with 3 symmetrical blue or purplish petals. Each filament is covered with fine hairs. This species, 6 to 12 in. tall, has irislike leaves and hairy pedicels and buds.

Virginia Spiderwort is occasional in partially shaded woods, roadsides, and other open places. Found throughout most of the eastern U.S., it is more common in the western half of our area.

Spiderworts are used as indicators of nuclear radiation; filament hairs change from blue to pink in response to low-level dosages. They are also cultivated as garden ornamentals.

Flowering: April, May.

Ohio Spiderwort *Tradescantia ohiensis* Raf.
Spiderwort Family Commelinaceae

In contrast to the Virginia Spiderwort (above), this plant is taller (2 to 3 ft.) and has smooth pedicels and buds.

The Ohio Spiderwort is common in meadows, edges of woods, and other open regions of the Midwest, but its range extends south to Florida and northeast to New England. It is less common in the mountainous portions of our area than further west.

The Zigzag Spiderwort, *T. subaspera,* is similar but has zigzag stems.

Flowering: May, June.

Jack-in-the-Pulpit

Slender Dayflower

Virginia Spiderwort

Ohio Spiderwort

Nodding Wild Onion
Allium cernuum Roth.

Lily Family
Liliaceae

This relative of the cultivated onion, *A. cepa,* is a non-weedy wild onion. The bent flower stalk (1 ft. tall) supports delicate pink (or occasionally white) flowers. The grasslike flat leaves are shorter than the flower stalk.

Found throughout the eastern U.S., the Nodding Wild Onion is often abundant in rocky soil and on dry slopes. In Kentucky and Tennessee it is more common in the western and middle portions than in the east. Local populations in the cedar glades in Georgia and Tennessee can be showy.

Other wild onions, including the Wild Onion (*A. stellatum*), Field Garlic (*A. vineale*), and Wild Garlic (*A. canadense*), lack the characteristic bent flower stalk.

Flowering: July-September.

False Solomon's-Seal
Smilacina racemosa (L.) Desf.

Lily Family
Liliaceae

The slightly zigzag stem of this plant is long (to 3 ft.) and arching. *Racemosa* refers to the inflorescence, composed of starlike flowers. A cluster of bright red berries, each dotted with purple, can be seen later in the summer.

Also called Solomon's Plume, it can be found in rich woods throughout our area and north to Canada.

The young shoots can be used like asparagus or added to salads.

Flowering: March-May.

Solomon's-Seal
Polygonatum biflorum (Walt.) Ell.

Lily Family
Liliaceae

The arching 3 to 4 ft. stem of this perennial bears 8 to 12 small (1/2 to 1 in.-long) creamy yellow bell-like flowers in the axils of the leaves. The berries are dark blue.

Look for this common plant along river banks and in rich woods or thickets. Its range includes all major parts of eastern North America from the Great Lakes to Florida.

The thick rootstocks may be boiled and eaten like potatoes. Medieval herbalists considered this plant a panacea, but modern research has not confirmed any supposed medicinal virtues.

Polygonatum pubescens, which shares the same common name, is a smaller plant with hairy (rather than smooth) leaves underneath.

Flowering: April, May.

Nodding Wild Onion

False Solomon's-Seal

Solomon's-Seal

Trout-Lily *Erythronium americanum* Ker.
Lily Family Liliaceae

Also called Dog-tooth Violet (though not related to violets), this plant produces flowers composed of 3 petals arranged inside 3 sepals. All 6 (tepals) are yellow with reddish-brown streaks. The 6 long stamens have anthers varying from yellow to cinnamon.

Large colonies of these wildflowers occur sporadically throughout much of the eastern and midwestern U.S. They are fairly common in rich woods throughout Kentucky and from the Central Basin of Tennessee eastward, especially at upper elevations of the Appalachians.

The origin of the name Trout-lily is uncertain. Is it because the mottled leaves suggest the appearance of trout, or because they flower at the time trout are biting?

The White Trout-lily, *E. albidum*, is similar but has white tepals. It is rare or occasional in Kentucky and in middle Tennessee.

Flowering: April, May.

Spanish-Bayonet *Yucca filamentosa* L.
Lily Family Liliaceae

Large white flowers, each with 3 sepals resembling the 3 petals, characterize this plant. The flowers form a panicle on a thick stalk often 5 to 6 ft. tall. Surrounding the base of the stalk is a rosette of thick, sharply pointed leaves, often with loose threads along their margins.

The original range is not known exactly; generally, it is considered native along the eastern coast from New Jersey south to Florida. In cedar glades and barrens, it is occasionally found along the edge of open areas where soils are deep enough for it to establish itself.

In pioneer days, fibers from the tough leaves were used for cord. The young flowers can be eaten in a salad. This and other species of *Yucca* are often planted as ornamentals.

Flowering: May, June.

False-Garlic *Nothoscordum bivalve* (L.) Britt.
Lily Family Liliaceae

This relative of garlic and onions (*Allium*) bears 3 to 10 small (1/2 in.-long) creamy yellow flowers in terminal umbels. The bulb and linear, basal leaves are odorless (thus the common name).

Found in fields and near granite and limestone outcrops throughout the eastern U.S. from the Midwest south to Alabama and Virginia, False-garlic is frequent in central and western Kentucky and in eastern Tennessee. It is common in the thin dry soils of cedar glades in Georgia and Tennessee.

This species can be distinguished from any *Allium* because the latter will have a strong odor and a perianth not as reflexed as that of False-garlic.

Flowering: April, May.

Trout-Lily

Spanish-Bayonet

False-Garlic

Toadshade

Trillium cuneatum Raf.

Lily Family

Liliaceae

Trilliums are triangular-shaped perennials with leaves, sepals, and petals in whorls of 3. In this species, the leaves are blotched, the sepals are green/reddish and semi-erect, and the petals are maroon, narrow, and erect.

The Toadshade is common in rich woods in our area. Although not found in open cedar glades, it often occurs (as seen here) in adjacent mixed hardwoods.

Other less common sessile-flowered (stalkless) trilliums appear in the Central South, including the Sessile Trillium, *T. sessile,* which has solid green leaves and wider petals.

Flowering: March-May.

Prairie Trillium

Trillium recurvatum Beck

Lily Family

Liliaceae

This trillium is similar to the Toadshade (above) except for the recurved (drooping) sepals.

The Prairie Trillium, despite its name, is common in rich woods. It occurs west of the Cumberland Plateau. E. Lucy Braun considers it an indicator species of the Western Mesophytic Forest Region (Introduction, Part 1).

Flowering: April, May.

Large-Flowered Trillium

Trillium grandiflorum (Michx.) Salisb.

Lily Family

Liliaceae

This common trillium, which grows to 18 in., is easily recognized by its stalked flowers, each with 3 white petals that turn pink as they age.

It typically forms colonies in rich woods over much of the eastern U.S.; its range extends as far south as northern Georgia and Alabama. It appears in eastern and southern Kentucky, in eastern Tennessee, and in the adjacent eastern Highland Rim portion of middle Tennessee.

Flowering: April, May.

Toadshade

Prairie Trillium

Large-Flowered Trillium

False Aloe

Daffodil Family

Manfreda virginica (L.) Rose
[*Agave virginica* L.]
Amaryllidaceae

This succulent perennial produces a rosette of leaves at the base of the tall (3 to 6 ft.) flowering stalk. The small (3/4 to 1 in.-long), inconspicuous, tubular flowers produce three-part capsules. When one brushes against the dried fruits with their loose seeds inside, a rattling sound results that is responsible for another common name, Rattlesnake-master.

This relative of the agaves of western deserts often occupies dry, rocky, especially calcareous places in the Midwest and occurs south to east Texas and Florida. In Kentucky, it is found especially in the Bluegrass Region. It is frequent and locally abundant in the cedar glades in Alabama, Georgia, and Tennessee.

Some botanists place *Agave*, along with *Yucca*, in the Agavaceae.

Flowering: June-August.

Star-Grass

Daffodil Family

Hypoxis hirsuta (L.) Coville
Amaryllidaceae

This short (3 to 6 in.) grasslike perennial has hirsute (hairy) leaves and small (3/4 in. across) flowers in clusters of three or more. The similar appearance of the sepals and petals gives the flower a starlike appearance.

Star-grass is widespread in southern Canada and throughout the eastern U.S. to northern Florida. It is fairly frequent in Kentucky except for the Bluegrass Region, where it is rare. It is found throughout Tennessee, but is especially frequent in the grassy zone of cedar glades.

Flowering: April, May.

Spider-Lily

Daffodil Family

Hymenocallis occidentalis (Le Conte) Kunth
[*H. caroliniana* (L.) Herb.]
Amaryllidaceae

This perennial is one of the more striking plants in our region. Like other members of the Daffodil Family, it grows from a large bulb. The leaves (all basal) are long and narrow. The flower stalk, which becomes 1 1/2 to 2 ft. tall, bears several flowers in an umbel.

Although other species of *Hymenocallis* are found only in the Coastal Plain region of the southeastern U.S., this one extends north to Tennessee and Kentucky, where it is infrequent in the western parts and rare or absent toward the east. The specimen seen here is one of a population along a stream bed in an open cedar glade in Tennessee.

Flowering: July, August.

False Aloe

Star-Grass

Spider-Lily

Blackberry-Lily *Belamcanda chinensis* (L.) DC

Iris Family Iridaceae

The common name of this tall (2 to 3 ft.) plant is misleading because it is related to neither the lily nor the blackberry; it is a member of the Iris Family. Its true identity is indicated by the flattened swordlike leaves. The flowers (1 1/2 to 1 3/4 in. across) resemble those of a lily except for 3 rather than 6 stamens. The fruits split open when ripe, exposing the black seeds that suggest blackberries (see inset).

This Asian plant has become naturalized over much of the eastern U.S. from New England and the Midwest to the Central South. In cedar glades in Alabama, Georgia, and Tennessee, it occurs typically in open glades along the edge of the shrub zone.

Flowering: June, July.

Crested Dwarf Iris *Iris cristata* Ait.

Iris Family Iridaceae

This woodland plant is like a miniature edition of the Bearded Iris, the official Tennessee cultivated flower. The showy flowers are borne by plants only 4 to 6 in. high. Each erect sepal has a small yellow or whitish fluted crest bordered with a white zone outlined in violet or purple.

The Crested Dwarf Iris, frequently found on wooded slopes and in ravines, appears from the District of Columbia to Missouri and south to the Piedmont of Georgia. It is scattered throughout Kentucky, except for the Bluegrass Region, and Tennessee, except for the Central Basin.

Flowering: April, May.

White Blue-Eyed Grass *Sisyrinchium albidum* Raf.

Iris Family Iridaceae

Blue-eyed grasses, *Sisyrinchium* species, grow in clumps, but the leaves are compressed into one plane as in irises. The small flowers include 3 sepals and 3 petals that look almost identical; each has a point on the end. In this species, the stems are 1/8 in. wide, and the flowers are pale blue to white.

This plant can be seen occasionally in dry woods and open places from New York to Wisconsin and south to Texas and Florida. It is found in Kentucky and Tennessee except for the high mountains and is conspicuous among the limestone glade flora in Alabama, Georgia, and Tennessee.

The similar Blue-eyed Grass, *S. angustifolium,* has somewhat wider leaves and bright blue flowers. It occurs in cedar glades but more often along roadsides.

Flowering: March-May.

Blackberry-Lily

Crested Dwarf Iris

White Blue-Eyed Grass

Yellow-Fringed Orchid
Orchid Family

Platanthera ciliaris (L.) Lindley
[*Habenaria ciliaris* (L.) R. Br.]
Orchidaceae

This showy perennial grows to 3 ft. but typically is half that height. Large lance-shaped leaves occur at the bottom of the flowering stalk; the leaves generally are smaller toward the top.

This species grows in wet acidic soil in either sun or partial shade. It is widespread in eastern North America from southern Canada to Florida and west to Texas. It is common in eastern Tennessee but infrequent in eastern and southern Kentucky and the Highland Rim of Tennessee.

A similar species with much the same distribution, the Crested Yellow Orchid, *P. cristata,* has smaller flowers with spurs shorter than the lips.

Flowering: June-September.

Lily-Leaved Twayblade
Orchid Family

Liparis lilifolia (L.) L.C. Rich.
Orchidaceae

The 4 to 10 in. raceme of this unusual plant typically bears 12 to 24 delicate semi-translucent pale purple flowers. Note the 2 glossy basal (2 to 5 in.) leaves that account for its common name.

The typical habitat is the moist banks of woodland streams (this one was growing by a pond). It is found from the northeastern U.S. to northern Georgia and west to Missouri. In Kentucky and Tennessee, it occurs throughout but only locally and infrequently.

Flowering: May, June.

Spring Coral-Root
Orchid Family

Corallorhiza wisteriana Conrad
Orchidaceae

This is one of a dozen or so saprophytic orchids of North America. Lacking chlorophyll, they absorb nutrients from soil humus. In this species, the slender flowering stalk (5 to 10 in. tall) bears 5 to 25 flowers, each with white petals spotted with magenta.

More common in the West, this unusual plant occurs rarely throughout the Southeast. Look for it along the banks of streams and in shady forests in central and western Kentucky and middle and eastern Tennessee.

The Autumn Coral-root, *C. odontorhiza,* also in our area (but absent in the Central Basin), flowers in August and September.

Flowering: April, May.

Grass-Pink Orchid

Calopogon tuberosus (L.) Britton, Sterns, and Poggenberg
Orchid Family
[*C. pulchellus* (Salisbury) R. Brown]
Orchidaceae

The 12 to 18 in. stem of this plant bears 3 to 25 deep pink (rarely white) flowers, each 1 to 1 1/2 in. across. The single long linear leaf is basal.

This orchid requires bright sunlight and moist acidic soil. Found in marshes, swamps, and bogs throughout eastern North America, it is fairly common in eastern and southern Kentucky, but infrequent in Tennessee from the eastern Highland Rim eastward.

Flowering: June.

Yellow-Fringed Orchid

Lily-Leaved Twayblade

Spring Coral-Root

Grass-Pink Orchid

Slender Ladies'-Tresses *Spiranthes gracilis* (Bigel.) Beck
Orchid Family Orchidaceae

In the genus *Spiranthes,* small whitish flowers are spirally arranged to form a spike. Of the approximately 300 species worldwide, about half a dozen are found in our area. In this one, several spikes may grow from a single root. The single row of flowers is strongly spiraled; the central portion of each flower is greenish.

Like other ladies'-tresses, this one is found in moist meadows. Primarily a Midwestern species, it occurs in our area mainly in eastern Tennessee and Kentucky.

In *S. cernua,* flowers are arranged in two or more rows; it is found (but infrequently) in grassy cedar glades.

Flowering: August, September.

Pink Moccasin Flower *Cypripedium acaule* Ait.
Orchid Family Orchidaceae

One of the 3 petals is distinctive among members of this family. In *Cypripedium,* the large odd petal accounts for the common names moccasin flower and lady's-slipper. Note the 2 (6 in.-long) basal leaves.

This wildflower is typically found in acidic soil in both wet and dry habitats. In our area, it appears on the Cumberland Plateau and less commonly on the eastern Highland Rim.

The rare Showy Lady's-slipper, *C. reginae,* is a larger northern species with a range extending into the eastern portion of our area. It has a pink and white pouch; also, its leaves are attached along the flower stalk.

Flowering: April-June.

Large Yellow Lady's-Slipper *Cypripedium calceolus* L. var. *pubescens* (Willd.) Correll
Orchid Family Orchidaceae

The "shoe" of this lady's-slipper is about 1 to 2 in. long. Note the leaves attached to the flower stalk. The brownish capsules above are the previous season's fruits from which minuscule seeds have already been released.

This relatively rare plant is scattered throughout much of our area except for the Central Basin. Requiring a rich acidic soil, it is usually seen on moist wooded slopes.

American Indians considered a boiled extract of roots from *Cypripedium* species to be effective as a sedative and as a treatment for various nervous diseases. Some such uses seem to be partially substantiated by current knowledge, but more scientific study is necessary before more definite statements can be made.

The natural urge to transplant members of the Orchid Family from the wild should be resisted; such efforts are rarely successful. Several kinds can be obtained from reliable specialty nurseries.

Flowering: April-June.

Slender Ladies'-Tresses

Pink Moccasin Flower

Large Yellow Lady's-Slipper

Part 2
Dicots

Lizard's-Tail *Saururus cernuus* L.

Lizard's-Tail Family Saururaceae

Obviously, the common name refers to the long (1 to 2 ft.), graceful spikes of small white flowers. The leaves are heart-shaped. The only member of its family in eastern North America, it spreads by means of rhizomes and may form extensive colonies.

This wildflower is common in swamps and along the margins of streams. Its range includes most of the eastern U.S. It is sold as an ornamental under the name of Swamp-lily.

Flowering: May-July.

Dock-Leaved Smartweed *Polygonum lapathifolium* L.

Buckwheat Family Polygonaceae

Smartweeds are so named because of the sharp taste of their leaves. The name knotweed is also applied because the stipules form cylindrical sheaths around the stem. An annual, this species reaches 3 ft. in height. The arching or drooping racemes as seen here make it easy to identify.

Dock-leaved Smartweed is common in disturbed, open low places. It is widespread in the eastern U.S. In the Central South, it is more common in the western portion. The leaves and seeds of many species of *Polygonum* have been used as a seasoning. The seeds provide food for many kinds of birds.

Flowering: May-October.

Pokeweed *Phytolacca americana* L.

Pokeweed Family Phytolaccaceae

This tall (to 10 ft.), coarse perennial is quite well known. The greenish-white flowers are arranged in erect spikes and the leaves are lanceolate. The purplish berries that form later have the same arrangement as the flowers.

Pokeweed is a common plant in disturbed habitats (especially fencerows) throughout most of eastern North America. In our area, it is more common in the east.

This plant is used as a popular potherb, poke sallet, especially in the southern U.S. Only tender young shoots should be used; the seeds, reddish stems, leaves, and roots are poisonous. In colonial times, the berries were a source of ink.

Flowering: July-October.

Lizard's-Tail

Dock-Leaved Smartweed

Pokeweed

Spring-Beauty *Claytonia virginica* L.
Purslane Family Portulacaceae

This family is characterized by flowers with 2 sepals, 5 petals, and 5 stamens. Note the pink-veined petals and linear leaves of this Spring-beauty. Since reproduction is mainly by root swellings called corms, plants typically occur in large masses.

Spring-beauty is found in various habitats, including rich deciduous meadows and city lawns. It can be seen throughout our area, west to Texas, and north to Canada.

The less common Carolina Spring-beauty, *C. caroliniana,* is similar but has wider, lanceolate leaves.

Because the corms of both species are sometimes boiled and eaten like small potatoes, the name Fairy Spuds is sometimes applied.

Flowering: March-May.

Limestone Fame Flower *Talinum calcaricum* Ware
Purslane Family Portulacaceae

This miniature perennial (4 to 6 in.) is easily overlooked even though it may occur locally in large numbers. The curved, somewhat cylindrical leaves are only 3/4 in. long; the flowers, borne in small clusters at the top of slender wiry stems, are less than 1/2 in. across. The flowers usually open only a few hours each afternoon.

Look for this floral gem in the thin rocky soil of the gravelly zone of Tennessee cedar glades.

Until 1967, botanists considered this plant to be *T. teretifolium,* which occurs on sandstone outcrops of the Cumberland Plateau. That year Stewart Ware described it as a new species.

Flowering: May-September.

Wild Ginger *Asarum canadense* L.
Birthwort Family Aristolochiaceae

Most apparent are the large cordate leaves arranged in pairs, but a close look reveals the solitary flowers at ground level attached to the stem at the fork of a pair of hairy-leaf petioles. Lacking petals, the 3 sepals, each with a pointed tip, form the cup-shaped calyx tube.

Wild Ginger occurs in rich soils in woods throughout the Central South, but seldom in large numbers. This photo was taken in Tennessee in a large sinkhole surrounded by glades.

Little Brown Jug, *A. arifolium,* also has cordate leaves and brownish flowers. However, the flowers are elongate and lack the pointed tips.

Wild Ginger is so called because the rootstocks are used as a seasoning similar to true ginger, which comes from an unrelated tropical plant.

Flowering: April, May.

Spring-Beauty

Limestone Fame Flower

Wild Ginger

Glade Sandwort

Arenaria patula Michx.

Pink Family

Caryophyllaceae

The more than 100 species of *Arenaria* are all known as sandworts. Well named, they are found in sandy soil where they form extensive mats. Also called Wild Baby's Breath, this small annual, 3 to 6 in. tall, has linear leaves. Note the 5 petals, each notched and about 1/2 in. long.

Although often associated with cedar glades, Glade Sandwort is widespread in limestone-based ecosystems from Texas to Ohio. It is frequent in the limestone soils in the Highland Rim and Bluegrass Regions.

Flowering: April-June.

Fire Pink

Silene virginica L.

Pink Family

Caryophyllaceae

Many species of *Silene* are called catchflies because crawling insects become stuck on their sticky stems, leaves, and flowers. The plant thus prevents these insects from interfering with the more desirable and effective pollination by flying insects. Fire Pink is a 1 to 2 ft.-tall catchfly with opposite, lanceolate leaves. Note the 5 notched petals at right angles to the corolla tube.

This brilliantly colored wildflower, common throughout our area, is most often seen on dry rocky banks or in open woods.

The Round-leaved Catchfly, *S. rotundifolia*, aptly described by both of these names, occurs in our area principally on sandstone cliffs of the Cumberland Plateau and adjacent Highland Rim.

Flowering: April, May.

Tall Buttercup

Ranunculus acris L.

Buttercup Family

Ranunculaceae

Although the name buttercup is often (but improperly) applied to daffodils, the genus *Ranunculus* includes the true buttercups. Like most buttercups, this species has 5 waxy yellow petals partially hiding 5 green sepals beneath, as well as numerous stamens and pistils spirally arranged in the center of the flower. The plant is typically 1 to 2 ft. tall and has leaves divided into 5 to 7 narrow, toothed leaflets.

This European native, naturalized throughout most of North America, is found most often in wet disturbed habitats.

Kidney-leaf Buttercup, *R. abortivus,* is a smaller (8 to 18 in. tall) but also common buttercup; its petals are less conspicuous and the basal leaves are undivided.

Flowering: April-July.

Glade Sandwort

Fire Pink

Tall Buttercup

Hooked Buttercup

Ranunculus recurvatus Poir.

Buttercup Family

Ranunculaceae

In comparison to the Tall Buttercup (preceding page), the Hooked Buttercup is a somewhat shorter (to 20 in.) plant with wider, 3-lobed leaves and smaller, less showy flowers. The sepals are recurved (angled downward); each fruit has a hooked style.

Except for the Central Basin, this wildflower is found throughout our area. It prefers moist soil.

Flowering: April, May.

Sharp-Lobed Hepatica

Hepatica acutiloba DC.

Buttercup Family

Ranunculaceae

Hepaticas grow low (2 to 5 in.) and have 6 to 10 petal-like sepals of white or tints of blue, pink, or lavender. Under them are 3 bracts (modified leaves that in this case resemble sepals). In *H. acutiloba,* the leaf lobes are pointed.

The Sharp-lobed Hepatica is fairly frequent throughout the Central South on rich slopes in mesophytic forests.

Flowering: March, April.

Golden-Seal

Hydrastis canadensis L.

Buttercup Family

Ranunculaceae

Six to 18 in. tall, this perennial plant produces several umbrellalike leaves, each with 5 lobes. Reddish buds open to form the feathery cream-colored flowers seen here. The fruits (berries) are also red. The common name is suggested by the yellowish rhizomes.

As a result of wholesale collecting, Golden-seal is no longer common. It is found in well-drained woods, more often in the eastern portion of our area and northward, especially in the mountains, to Vermont.

Flowering: April, May.

Glade Larkspur

Delphinium carolinianum Walt.
(undescribed subsp.)

Buttercup Family

Ranunculaceae

Larkspur refers to the single spur formed by 1 of the 5 sepals (resembling petals). The 4 small petals at the center are less conspicuous. In this species, the 3/4 in.-long flowers, mostly white or sometimes with pink spurs, are clustered along the upper portion of the stem (3 to 4 ft.). The deeply dissected leaves are at the base of the stem.

Delphinium carolinianum occurs occasionally in open dry woods, fields, and rocky or sandy soils. Its range includes an area west of our region (Texas, Missouri) and also the Piedmont of Georgia and South Carolina. The glade subspecies featured here, recognized only recently (formerly called *D. virescens*), is common in cedar glades and other rocky habitats in the Central Basin.

Spring (or Dwarf) Larkspur, *D. tricorne*, is a shorter (12 to 18 in.) plant with deep blue flowers. It is found in woods throughout most of our area.

Flowering: May-July.

Hooked Buttercup

Sharp-Lobed Hepatica

Golden-Seal

Glade Larkspur

Rue-Anemone

Anemonella thalictroides (L.) Spach
[*Thalictrum thalictroides* (L.) Boivin.]

Buttercup Family

Ranunculaceae

This small (6 in.-tall), dainty perennial grows from tuberous roots. Its "petals" are actually sepals; the number varies from 5 to 10, and the color from white to pinkish. Flowers are about 1/2 to 1 in. across. Petals are absent. The leaves are whorled, compound, and 3-lobed.

Rue-anemone is common throughout much of the eastern U.S. In cedar glades, it is found along the edges and in clearings of glade woods.

False Rue-anemone, *Isopyrum biternatum,* resembles the plant described above but is taller (8 to 16 in.) and has smaller flowers. Also, the leaves are alternate, unlike those of Rue-anemone. It can be found in cedar glades and in much of the Highland Rim and Bluegrass Basin.

Flowering: April.

Wild Columbine

Aquilegia canadensis Pursh

Buttercup Family

Ranunculaceae

The graceful inverted flowers of columbine have 5 spurs, each an extension of a petal. The nectar inside the spurs attracts pollinators with tongues long enough to reach it. There are also 5 petal-like sepals. In *A. canadensis,* the coral/cream flowers are 1 1/2 to 2 in. long. The only columbine native to the eastern U.S., it is quite unmistakable.

Wild Columbine is rather common, especially on dry, partially shaded banks and rock outcrops, throughout much of the eastern U.S. Frequently found in calcareous soil, it thrives in cedar glade woods.

Hybrids of various columbines native to the western U.S., having larger flowers in many colors, are cultivated in rock gardens.

Flowering: April-June.

Leather-Flower

Clematis viorna L.

Buttercup Family

Ranunculaceae

This common name is applied to several perennial *Clematis* vines of our area. Each of the urn-shaped flowers is enclosed by 5 thick, dull red-to-pinkish petal-like sepals. Petals are absent. In this species, the flowers are almost 1 in. long. The fruits that follow are plumed. Each of the opposite leaves has 3 to 7 leaflets.

This unusual vine is common along the edge of woods and thickets throughout the Central South and adjacent states.

Other leather-flowers of our area: *C. glaucophylla, C. versicolor, C. pitheri,* and *C. reticulata.* To distinguish between them, consult a wildflower guide by Duncan and Foote or Wharton and Barbour (Bibliography).

Flowering: May-June.

Rue-Anemone

Wild Columbine

Leather-Flower

May-Apple

Podophyllum peltatum L.

Barberry Family

Berberidaceae

May-apples have shiny umbrellalike leaves. Their solitary flowers are in the axils of paired leaves. Each cuplike flower, 1 to 2 in. wide, has 6 to 9 waxy petals and twice that number of stamens.

May-apples are found in openings in moist woods, invariably in colonies containing dozens of plants. This wildflower, documented throughout most of our area, is less common in the Central Basin than elsewhere.

American Indians valued the May-apple as a laxative and a treatment for warts. At present, the drug industry is using quantities of the rhizomes as anti-cancer medicinals. Dr. William Meyer, botanist at the University of Kentucky, has proposed that May-apple could be cultivated much like Ginseng as a cash crop. All parts of the plant, except for the edible ripe fruits, are poisonous.

Flowering: April, May.

Twin-Leaf

Jeffersonia diphylla (L.) Pers.

Barberry Family

Berberidaceae

The flowers and leaves of this plant are on separate stalks, both arising from the rhizome. Each flower is about 1 in. across and has 8 petals. Both the common name and the specific epithet aptly describe the unusual leaves.

The distribution pattern of Twin-leaf within our area is unusual. Wharton and Barbour report it as occurring in Kentucky, mainly in the limestone Bluegrass Region; in Tennessee and Alabama it is associated with the rich moist woods in the Highland Rim. In general, it is rare to infrequent.

C. H. Persoon, a botanist, named the genus in honor of his friend Thomas Jefferson.

Flowering: March, April.

Dutchman's Breeches

Dicentra cucullaria L. Bernh.

Poppy Family

Papaveraceae

Highly dissected fernlike leaves and distinctly shaped white flowers are characteristic of *Dicentra.* In this species, each flower, about 1/2 in. long, has 2 inflated spurs resulting in the inverted pantaloon appearance.

Dutchman's Breeches occurs in moist but well-drained rich woods. It is seen throughout most of the Central South but is more common in the eastern part.

Squirrel-corn, *D. canadensis,* has almost identical foliage, but the flowers are fragrant, more heart-shaped, and lack prominent spurs.

Flowering: April.

May-Apple

Twin-Leaf

Dutchman's Breeches

Duck River Bladderpod

Lesquerella densipila Rollins

Mustard Family

Brassicaceae

The Mustard Family, Brassicaceae or Cruciferae, is well represented in our area and other temperate regions. The crucifer refers to the 4 petals arranged to form a cross.

In the early 1950s, Dr. Reed Rollins of Harvard University studied two mustard genera, *Lesquerella* and *Leavenworthia*, focusing particularly on representatives in the Interior Low Plateau. The wildflower featured here is one of several new species he described. Bladderpod refers to the swollen rounded seedpods, a feature of the genus. Dense, short hairs on the pods help distinguish this from other bladderpods. The Duck River Bladderpod stands 6 to 10 in. tall.

This mustard appears sometimes in glades but more often in fields where it may form large masses. Found principally in the drainage basin of the Duck River, it is endemic to the Central Basin.

Flowering: April, May.

Stone's River Bladderpod

Lesquerella stonensis Rollins

Mustard Family

Brassicaceae

Seen here are the distinctive white flowers of the Stone's River Bladderpod; it is known only from along the east fork of the Stone's River in Rutherford County, Tennessee. One population is now managed as a State Natural Area to insure its survival.

Other endemic bladderpods in our area include *L. perforata,* another white-flowered species known only from the Spring Creek area, Wilson County, Tennessee; *L. lyrata,* cedar glades in northern Alabama; and *L. lescurii,* Central Basin.

The seed-oil from several of the many western *Lesquerella* species is used as an industrial lubricant .

Flowering: April, May.

Garlic-Mustard

Alliaria officinalis Andrz.

Mustard Family

Brassicaceae

This 1 to 3 ft.-tall plant has simple alternate leaves with large teeth along their margins. Each of the 4 rounded petals is about 1/4 in. long. The slender pods, called siliques, differ from Pea Family pods because of the thin partition separating the two seed chambers.

Introduced from Europe, this biennial has spread throughout most of the eastern U.S. Like other weeds, it thrives in disturbed soil but tolerates shade better than others.

As suggested by its common name, it has the odor of garlic; it is sometimes used as a potherb. The specific epithet *officinalis* indicates that it was probably used as a medicinal, but its particular efficacy, if any, appears to be lost in antiquity.

Flowering: April, May.

Duck River Bladderpod

Stone's River Bladderpod

Garlic-Mustard

Nashville Mustard *Leavenworthia stylosa* Gray
Mustard Family Brassicaceae

Not widely known, wildflowers of the genus *Leavenworthia* are well represented in the Central Basin, especially in cedar glades. The fragrant flowers are solitary on stalks separate from the basal leaves; the petals are notched. The fruits are elongated seedpods (siliques). The two color forms of Nashville Mustard are usually separated geographically, but they occasionally grow together as seen here. Populations in the Nashville and Lebanon areas are generally yellow-flowered, whereas those around Cedars of Lebanon State Park and Murfreesboro have white corollas with yellow centers. Sometimes the petals are violet instead of white in the latter form. The small rounded leaves lie nearly flat on the ground below the flowers borne on stalks only 2 to 3 in. long. The specific epithet *stylosa* refers to the prominent style projecting at the end of each fruit.

This species is endemic to glades in the Central Basin, where it forms large, showy masses in spring.

Flowering: February-May.

Glade-Cress *Leavenworthia uniflora* (Michx.) Britt.
Mustard Family Brassicaceae

In contrast to the previous species of *Leavenworthia*, this Glade-cress has fruits that lack the extended style and more dissected leaves. Also, it typically occurs in smaller populations of scattered plants rather than in masses.

Frequent in cedar glades, it is also found in other limestone habitats in the mid-continent.

Altogether, 12 taxa (species and/or varieties) of *Leavenworthia* are considered to be cedar glade endemics. See the glade checklists and the autecological treatment in the Baskins's 1989 symposium article.

Flowering: March-May.

Cut-Leaved Toothwort *Dentaria laciniata* Muhl.
Mustard Family Brassicaceae

Toothworts *(Dentaria)* are low, early-flowering plants with racemes of 4-petaled white-to-pinkish petals. About 1 ft. tall, this species has deeply cut, toothed leaves arranged 3 per node.

It is common in woods throughout the Central South and north to New England and southern Canada.

Other toothworts in our area include *D. diphylla*, with paired leaves divided into 3 broad leaflets, and *D. multifida*, with leaves more finely divided than those of *D. laciniata.*

The pungent roots have been chewed to alleviate toothaches. Lee Peterson recommends them as spicy additions to salads.

Flowering: March, April.

Nashville Mustard

Glade-Cress

Cut-Leaved Toothwort

Lime Stonecrop *Sedum pulchellum* Michx.
Orpine Family Crassulaceae

Stonecrops in our area are low-growing plants with succulent leaves. The tiny white-to-pink flowers of Lime Stonecrop are arranged in inflorescences with 3 to 7 curved branches. The leaves (about 1/2 in.) are narrow and cylindrical.

The plants usually form a mat on thin soil over limestone bedrock, thus the name stonecrop. Sometimes seen on the Highland Rim, it is more common in the calcareous Bluegrass Region and the Central Basin, especially in cedar glades.

Woodland Stonecrop, *S. ternatum*, has white flowers and flattened round leaves in threes on the stems.

Flowering: May, June.

Early Saxifrage *Saxifraga virginiensis* Michx.
Saxifrage Family Saxifragaceae

Rosettes of light green leaves surround the foot-long stalks bearing panicles of small white flowers. Each flower is about 1/4 in. across and has 5 sepals, 5 petals, and 10 stamens.

This wildflower is found on partially shaded limestone cliffs, usually in thin soil, in the Bluegrass and Central Basin Regions.

Foamflower, *Tiarella cordifolia*, also in the Saxifrage Family, looks similar; however, its tiny white flowers form denser racemes, and its leaves are larger and shaped like those of maple.

Flowering: March, April.

Sulphur Cinquefoil *Potentilla recta* L.
Rose Family Rosaceae

Typical of the Rose Family are the flowers, each with 5 sepals, 5 rounded petals, and numerous stamens and pistils occupying a cuplike depression in the center of the flower. The leaves are alternate, often compound, with prominent stipules. In our area, many are woody plants (e.g., blackberries, wild roses, and black cherry trees).

Cinquefoil means "5 leaves," referring to some of the common species with 5 leaflets; actually, cinquefoils may have 3 to 9 leaflets per leaf. Their flowers are typical of the family; the petals are yellow in *P. recta*, a European weed (2 ft.). The 5 to 7 narrow leaflets are prominently toothed; the flowers are about 3/4 in. across.

Also called Rough-fruited Cinquefoil, it is common along roadsides and in fields from Canada to Tennessee; its distribution is expanding.

Dwarf Cinquefoil, *P. canadensis,* is a much smaller plant that grows prostrate like a strawberry plant. It has 3 to 5 leaflets and brighter yellow petals.

Flowering: May-July.

Lime Stonecrop

Early Saxifrage

Sulphur Cinquefoil

Prairie Mimosa *Desmanthus illinoensis* (Michx.) MacM.
Pea Family Fabaceae

This bushy (3 to 4 ft.) sometimes prostrate plant has doubly pinnate leaves, each with many tiny leaflets. The stalked flower heads produce curved pods often twisted together in round clusters.

Found from South Carolina and north Georgia to the prairie states, Prairie Mimosa occurs throughout our area infrequently and sporadically. Its habitat varies from dry sandy soil to river banks.

The Sensitive Brier, *Schrankia microphylla,* has a similar distribution. However, it has larger pink flowers and linear seedpods with hooked prickles.

Flowering: June, July.

Goat's-Rue *Tephrosia virginiana* (L.) Persoon
Pea Family Fabaceae

The large bicolored flowers and compound leaves easily identify this 1 to 2 ft. plant. Both leaves and stem are covered with silky hairs.

This plant is common along roadsides and dry openings in oak and pine woodlands from Florida north to Canada and west to Texas. It is found throughout our area except in the limestone soils of the Central Basin and the Bluegrass Region.

The roots contain the poisonous compound rotenone. North American Indians used them as an insecticide and fish poison.

Flowering: June, July.

Nashville Breadroot *Pediomelum subacaulis* (Torrey and Gray) Rydberh
Pea Family [*Psoralea subacaulis* (Torrey and Gray)]
Fabaceae

Though sometimes mistaken for a lupine *(Lupinus)*, this showy wildflower, usually about 6 in. tall, is easily identified by its palmately compound leaves and dense racemes of purplish-blue flowers. Nashville Breadroot is found mainly in the grassy zone of the cedar glades in Tennessee, Alabama, and Georgia. Its masses of flowers make it quite spectacular.

The nutritious tuber of the related Prairie Turnip, *P. esculenta,* was used as a subsistence food by 19th-century pioneers crossing the American prairie. The root of this glade species may have been used in a similar fashion.

The related Sampson's Snakeroot, *P. psoraloides,* is a tall (1 to 2 ft.), slender plant with bluish flowers. It is found sporadically in dry open woods throughout much of our area.

Flowering: April, May.

Prairie Mimosa

Goat's-Rue

Nashville Breadroot

Blue False Indigo *Baptisia australis* (L.) R. Br.
Pea Family Fabaceae

This rather conspicuous plant (3 to 4 ft. tall) has large cloverlike leaves up to 3 in. long. The dark blue-violet flowers (about 1 in.) are arranged into racemes.

Gattinger found this species in "cedar glades at Lavergne [Tenn.]." It is now known to be also in glades in Georgia and in moist habitats in the Highland Rim, such as wet prairies.

Two somewhat similar *Baptisia* species are Wild Indigo, *B. tinctoria,* and White False Indigo, *B. leucantha.* The former has yellow flowers and the latter white. Both are widespread and prefer dry open habitats.

The legume that is the source of the blue dye indigo is *Indigofera tinctoria,* an Old World legume with tiny reddish flowers.

Flowering: May, June.

Tennessee Milk-Vetch *Astragalus tennesseensis* Gray
Pea Family Fabaceae

This is perhaps the most common of the *Astragalus* species found in our area. The hairy-pinnate leaves and pealike flowers are characteristic; the cream-colored flowers form a raceme about 2 to 3 in. long.

The Tennessee Milk-vetch often occurs at the edge of cedar glades. Gray's manual lists it as occurring from "southern Ill. to Ala."

At least two other species of *Astragalus* are known in our area: Rattle-vetch, *A. canadensis,* with yellow flowers; and Guthrie's Ground Plum, *A. bibullatus,* which has bluish-purple flowers. The latter is known only from Rutherford County, Tennessee.

Of the 375 or more species (mostly western) of *Astragalus,* many accumulate the element selenium, causing a type of poisoning in livestock known as loco disease.

Flowering: April, May.

Gattinger's Prairie-Clover *Dalea gattingeri* (Heller) Barneby
Pea Family [*Petalostemum gattingeri* Heller]
Fabaceae

This low-growing glade plant has 5 to 7 leaflets on each fine, narrow leaf and dense, elongated rose-purplish (rarely white) flower heads on wiry reddish stems.

Often a dominant ground cover, this endemic of the Southeast occurs typically in the thin soil in open cedar glades in middle Tennessee, northern Alabama, and northwestern Georgia.

A similar species, *D. purpureum,* known as Purple-tassels, is more upright and differs in certain technical features. Duncan and Foote state in their *Wildflowers of the Southeastern United States,* "The bracts at the base of the flower heads are acuminate from narrower bases and the calyx is appressed-hairy." Rare in our area, it extends westward to New Mexico and northward to southern Manitoba. Look for it in prairie remnants.

Flowering: May, June.

Blue False Indigo

Tennessee Milk-Vetch

Gattinger's Prairie-Clover

Price's Wood-Sorrel — *Oxalis priceae* Small subsp. *priceae*
Wood-sorrel Family — Oxalidaceae

This showy perennial has leaves typical of the genus; each consists of 3 bilobed leaflets. The flowers are at least 1/2 in. across.

Although sometimes seen in woods, it is more characteristic of cedar glades, where it may be abundant. It is endemic to the southeastern U.S.

A more widespread relative is the Yellow Wood-sorrel, *O. stricta*. Its flowers are smaller and lack the distinctive red spots at the base of the petals.

The family name indicates the presence of oxalic acid. Although the leaves are recommended as a tart addition to salads, too much can be toxic.

Flowering: April, May.

Violet Wood-Sorrel — *Oxalis violacea* L.
Wood-sorrel Family — Oxalidaceae

Though this plant resembles the previous one, it is somewhat smaller, more delicate, and has violet flowers. The leaves contain the same violet pigment, especially noticeable on their undersides.

The Violet Wood-sorrel occurs both in woods and in open places, including the grassy zone of cedar glades. Although generally not common, it occurs throughout most of the eastern U.S.

Flowering: March-May.

Wild Geranium — *Geranium maculatum* L.
Geranium Family — Geraniaceae

This is the largest (to 2 ft.) and most showy of our native geraniums. The term Cranesbill is often applied to the genus because the fruit ends in a long beak. Note the 5-lobed hairy leaves.

The Wild Geranium is a perennial found in rich woods throughout much of our area, but it is more common in the mountains in eastern Alabama, Tennessee, and Kentucky.

Several other Cranesbills, particularly the annual Carolina Cranesbill, *G. carolinianum*, occur as weeds in disturbed places.

The common potted Geranium belongs in the same family but is of South African origin and has been assigned to the genus *Pelargonium*.

Flowering: April-June.

Price's Wood-Sorrel

Violet Wood-Sorrel

Wild Geranium

Rosy Milkwort

Polygala cruciata L.

Milkwort Family

Polygalaceae

This genus is represented by about a dozen species in our area. Most have cloverlike heads as seen here, but this one is the showiest. In addition to the deep pink flowers, this species is recognized by the narrow leaves arranged 4 to a whorl in the form of a cross.

This milkwort favors seasonally moist habitats, such as May Prairie, a wet prairie in the eastern Highland Rim, where this specimen was found. It has a wide distribution from Florida to Texas, north to Nebraska and Maine. Within our area, it is found on the Cumberland Plateau and adjacent Highland Rim.

The Field Milkwort, *P. sanguinea*, has similar flowers but leaves attached one per node. It is found sporadically throughout Kentucky and Tennessee.

Flowering: July-October.

Flowering Spurge

Euphorbia corollata L.

Spurge Family

Euphorbiaceae

Break the stem or a leaf of this plant or other members of the family and you will notice a milky sap. What appears to be a flower is actually a cluster of tiny imperfect flowers surrounded by 5 petal-like white bracts.

This common plant occurs widely throughout the eastern U.S., usually in dry disturbed soil. The glade checklist (Bibliography) indicates its presence in the cedar glades of Alabama, Tennessee, and Georgia.

Plants of the *Euphorbia* are poisonous if ingested. Nevertheless, this spurge is sometimes used by florists in arrangements.

Flowering: May-September.

Prairie-Tea

Croton monanthogynus Michx.

Spurge Family

Euphorbiaceae

The photo makes it clear why this unspectacular plant is not included in most wildflower books. Its leaves are silvery green because of dense stellate hairs visible with a hand lens. As in the preceding species, the flowers are greatly reduced in size, but there are no bracts to make them showy.

The habitat and the distribution are much like those of the Flowering Spurge. It is commonly found in cedar glades with its close relative Woolly Croton, *C. capitatus*, which has 3 bifid styles instead of 2 as in Prairie-tea.

Flowering: June-October.

Rosy Milkwort

Flowering Spurge

Prairie-Tea

Spotted Jewelweed

Impatiens capensis Meerburg

Touch-me-not Family

Balsaminaceae

The pendant, brightly colored flowers (1 in. long) are typical of jewelweeds. Another common name, touch-me-not, refers to the ripe fruit pods that burst open when touched, an effective mechanism for dispersing seeds.

Jewelweeds are common in shady, moist habitats, especially along stream banks. They are widely distributed throughout our area west to Missouri and north to Newfoundland.

The Pale Jewelweed, *I. pallida,* has pale yellow flowers. The juice from the leaves of both species is used to counteract the irritation of poison ivy.

Flowering: July, August.

Common St. John's-Wort

Hypericum perforatum L.

St. John's-wort Family

Guttiferae

St. John's-worts are herbs or shrubs with opposite leaves and flowers with 5, usually yellow, petals and numerous stamens. Common St. John's-wort is 1 1/2 to 2 1/2 ft. tall, highly branched, with narrow leaves and numerous flowers 1/2 to 3/4 in. across. The petals are black-dotted near the margin.

The only alien *Hypericum* species in our region, it grows as a weed in open areas such as roadsides, fields, and other disturbed places.

Two other species of *Hypericum,* both found in the cedar glades, are *H. sphaerocarpum,* which is similar but has wider leaves, and *H. dolabriforme,* which has longer leaves and larger (1 in. across) flowers. The variety *turgidum* of *H. sphaerocarpum* occurs in middle Tennessee glades; *H. dolabriforme* is found in the glades and barrens of east Tennessee.

Flowering: July, August.

Swamp Rose-Mallow

Hibiscus moscheutos L.

Mallow Family

Malvaceae

Although considered tropical plants, 19 species of *Hibiscus* occur in North America. Swamp Rose-mallow is perhaps the largest (to 7 ft.) and most conspicuous one in our region. A distinguishing feature of the family is that the stamens are united into a column surrounding the style. In most cases, the petals of this species are white or cream with a crimson center.

Found in marshy ground throughout our area, this plant is more common in the western portion.

Rose-mallow, *H. militaris,* is a smaller plant with light pink petals. Other, perhaps better known, members of the family are cotton, okra, and hollyhock.

Flowering: August, September.

Spotted Jewelweed

Common St. John's-Wort

Swamp Rose-Mallow

Confederate Violet

Viola papilionacea Pursh var. *priceana*

Violet Family — Violaceae

This is one of several variants of the species known generally as the Common Blue Violet. The leaves are heart-shaped, coarse, and veiny. The flower is usually blue-violet, but the Confederate Violet has grayish-white flowers with violet veining.

Confederate Violet occurs in large colonies in meadows, thickets, and lawns (where it may become weedy). Both the species as a whole and the Confederate variety are found throughout the eastern U.S., as well as in our area.

Flowering: April, May.

Canada Violet

Viola canadensis L.

Violet Family — Violaceae

Unlike the Confederate Violet (above), the whitish flowers and leaves of this 8 to 16 in.-tall plant are on the same stalk. The mostly white petals have a yellow eyespot with brown-purple veins at their base; they are purple-tinged on the back.

This woodland violet ranges from southern Canada to the Cumberland Plateau and adjacent Highland Rim of our area.

Other violets with white flowers include White Violet, *V. striata,* which has cream or ivory colored petals, the lateral ones strongly bearded, and deeply cut stipules; and Field Pansy, *V. rafinesquii,* a weedy alien with tiny whitish or pale blue pansylike flowers and spoon-shaped leaves with dissected leafy stipules at their bases.

Flowering: April, May.

Eggleston's Violet

Viola egglestoni Brainerd

Violet Family — Violaceae

Named for the botanist Willard W. Eggleston (1863-1935), this dark blue violet is distinguishable from other blue violets because of its highly variable, often deeply lobed leaves.

Once thought to be limited to cedar glades in Tennessee and Kentucky, it is now known to exist in adjacent portions of Alabama, Georgia, and Indiana.

Flowering: April, May.

Confederate Violet

Canada Violet

Eggleston's Violet

Birdfoot Violet *Viola pedata* L.
Violet Family Violaceae

This violet has large (1 to 1 1/4 in. diameter) flowers and deeply divided leaves. The 2 upper petals are usually darker than the other 3, and the stamens are orange-tipped. The divided leaves account for the specific epithet *pedata*.

Normally found in sandy soils, this violet grows in open woods and dry slopes throughout our area except for the Bluegrass Region and Central Basin.

Flowering: April, May.

Passion Flower *Passiflora incarnata* L.
Passion Flower Family Passifloraceae

The unusual flowers, 1 1/2 to 2 in. across, are followed by yellowish fruits the size of lemons. The name wild apricot or maypop is often applied to the fruits that have sweet, edible flesh around the seeds and pop loudly when stepped on.

Though it has a beautiful flower, the plant grows as a weedy vine in fencerows and other disturbed, sunny locations. It is common throughout the Southeast, north to southern Ohio, and west to Oklahoma.

This plant was selected by the schoolchildren of Tennessee as the official state wildflower (the official cultivated flower is the Iris).

The Yellow Passion Flower, *P. lutea*, has much smaller, yellow flowers and blackish fruits. Its habitat and distribution are similar to those of *P. incarnata,* but it is less common.

Flowering: May-September.

Prickly-Pear *Opuntia humifusa* (Raf.) Raf.
Cactus Family [*O. compressa* (Salisb.) Macbr.]
[*O. rafinesquii* Engelm.]
Cactaceae

This is the only cactus native to our region. The flattened green stems or pads are specialized for both water storage and photosynthesis. The protective spines are actually greatly reduced leaves. The smaller, more numerous bristles are called glochids.

This species is found in open sandy places throughout the Southeast and extends north to Massachusetts and west to Oklahoma.

The pulp of the reddish fruits or pears (August-October) can be eaten, made into jelly, or used to make a pleasing cold drink. Also, the leaf pads can be cooked as a vegetable after removing the skin and spines.

Flowering: May, June.

Birdfoot Violet

Passion Flower

Prickly-Pear

Primrose-Willow *Ludwigia peploides* (H.B.K.) Raven
Evening-primrose Family [*Jussiaea decurrens* (Walt.) DC]
Onagraceae

Also called water-primroses, *Ludwigia* species are mostly tropical. The stems of this aquatic perennial produce roots at their lower nodes. Note the long-stalked, shiny flowers (3/4 to 1 in. across), each with 5 petals.

This somewhat weedy wildflower grows in shallow water or in damp soil of marshes, sloughs, and mudflats throughout our area and west to Texas and Mexico.

Another species, *L. decurrens,* also called Primrose-willow, is an annual (sometimes over 3 ft.). It has stems with narrow wing angles on the internodes and flowers with only 4 petals.

Flowering: June-September.

Missouri Evening-Primrose *Oenothera missouriensis* Sims
Evening-primrose Family Onagraceae

Most evening-primroses are perennials found in prairies or open rocky places. They have a distinctive flower: 4 petals attached at the end of a long calyx tube. The stigma has 4 branches forming a cross. The common name refers to the opening of the flowers in late afternoon. This species has large (3 to 4 in. across) flowers produced on a plant less than 1 ft. in height.

Manuals give its range as Missouri to Texas, but this specimen was photographed at a disturbed cedar glade near Murfreesboro, Tennessee. Considered an endangered species in Tennessee, it is protected by law.

Several other evening-primroses have yellow flowers, but they are smaller than those of *O. missouriensis.* The Common Evening-primrose, *O. biennis,* has several flowers at the top of a 3 to 5 ft.-tall stem. The three-lobed Evening-primrose, *O. triloba,* is a shorter, and less common, plant with divided leaves.

Flowering: May, June.

Showy Evening-Primrose *Oenothera speciosa* Nutt.
Evening-primrose Family Onagraceae

The only pink-flowered (or sometimes nearly white) evening-primrose in our area, it has typical features of the genus, which can be seen in the flowers (see the previous description for details).

Sometimes called the White Evening-primrose, it is believed to be native to the prairies of North America. Since European settlement, it has been spread by cultivation and become naturalized throughout our area and other southern states. It often carpets the edges of highways.

Flowering: May, June.

Primrose-Willow

Missouri Evening-Primrose

Showy Evening-Primrose

Ginseng — *Panax quinquefolius* L.

Ginseng Family — Araliaceae

This 8 to 15 in.-tall herbaceous perennial produces 1 to 3 compound leaves, each with 5 leaflets. The radiating cluster of tiny greenish flowers is not nearly as showy as the fall berries seen here.

Until two centuries ago, Ginseng was common in rich deciduous forests throughout eastern North America. As a result of widespread collecting, it has become relatively rare. It is more frequent in the eastern portion of our range than in the west. Collectors dig and dry the "sang" roots, which are shipped to the Orient for use in panaceas and aphrodisiacs.

Dwarf Ginseng, *P. trifolium*, is a smaller plant (to 6 in.) with whitish flowers and yellow berries. Also occurring in rich woods, it is found in the mountains of Tennessee and Kentucky, including the eastern portion of the Central South.

Flowering: June, July.

Poison Hemlock — *Conium maculatum* L.

Parsley Family — Apiaceae

This tall (to 8 ft.), coarse biennial, introduced from Eurasia, has become a weed. Note the small white flowers in umbels and the finely divided leaves. The purple spots on the blue-gray stem suggested the specific epithet *maculatum* (spotted) to Linnaeus.

Like many other weeds, Poison Hemlock is found in open disturbed areas and is widespread throughout the eastern U.S.

The sap of the plant is quite poisonous if ingested. In ancient Greece, it was used to execute criminals and troublemakers like Socrates.

Flowering: May, June.

Shooting-Star — *Dodecatheon meadia* L.

Primrose Family — Primulaceae

The unusual flowers of Shooting-star form an umbel at the top of the 15 to 25 in. stalk. Smooth oval leaves form a rosette at its base. The 5 swept-back petals, which are sometimes pink, and the 5 united stamens contribute to the rocket appearance.

Although widespread and sometimes locally abundant, the Shooting-star is not a common plant. It ranges from northern Alabama and Georgia west to Texas and north to Pennsylvania, and occurs generally in somewhat alkaline (basic) soils. Since it inhabits both open sunny areas (including glades) and shaded woods, some recognize the existence of two ecotypes (i.e., ecological variants adapted to local conditions) within the species.

Flowering: March-May.

Ginseng

Poison Hemlock

Shooting-Star

Indian-Pink
Spigelia marilandica L.

Logania Family
Loganiaceae

This striking wildflower grows 1 to 2 ft. tall, usually in rich woods. Each plant has 4 to 7 pairs of opposite sessile leaves and a curving spike of upright flowers (1 in. long). The corolla is coral red outside and a pure yellow inside. The coloration seen here is partly the result of the eerie light produced by a nearly total annular eclipse of the sun on May 30, 1984.

The range includes our area and extends north to Maryland and Indiana, west to Texas, and south to northern Florida.

An extract from this plant, also known as Pink-root, has been used to get rid of intestinal parasites. Care should be exercised because the effective alkaloid, spigeline, is poisonous.

Flowering: May, June.

American Columbo
Swertia caroliniensis (Walt.) O. Ktze

Gentian Family
Gentianaceae

Though the genus *Swertia* includes numerous tall, showy herbs of Eurasia, Africa, and North America, this is the only representative of the group in the eastern U.S. A triennial, Columbo produces a rosette of large (1 ft.-long) smooth leaves for two growing seasons before producing a panicle in the third year. The flowers, each about 1 in. across, are greenish with purple spots.

This relatively rare plant is widespread in the eastern U.S. Preferring limestone soil, it grows on roadsides and in glade woods in the Central Basin more often than in other parts of the Central South.

The plants at lower right are Yellow Sweet-clover, *Melilotus officinalis*. Both it and White Sweet-clover, *M. alba,* are common roadside weeds.

Flowering: May, June.

Rose-Pink
Sabatia angularis (L.) Pursh

Gentian Family
Gentianaceae

This multi-branched, 1 to 3 ft. annual has wing-angled stems. The fragrant flowers vary in color from the usual pink seen here to almost white, but the greenish star at the center helps identify them as a *Sabatia* flower.

This attractive plant prefers moist meadows but also may be in fencerows and occasionally in open glade woods. A plant of the eastern U.S., it is less common in the Bluegrass Region of Kentucky and in western Tennessee than in other parts of our area.

Also called Rose-pink, *S. brachiata* is similar but has narrow leaves that taper to a base.

Flowering: July, August.

Indian-Pink

American Columbo

Rose-Pink

Periwinkle *Vinca minor* L.

Dogbane Family Apocynaceae

Also called Myrtle, this evergreen herbaceous vine makes an attractive ground cover. Note the flowers (1 in. across) with 5 asymmetrical petals that give it a pinwheel appearance.

A native of Europe, Periwinkle is seen along partially shaded roadsides and in cemeteries where it has escaped cultivation.

Flowering: March-June.

Blue Dogbane *Amsonia tabernaemontana* Walt.

Dogbane Family Apocynaceae

Also called Blue-star, this and other species of *Amsonia* are herbs with alternate leaves and pale blue starlike flowers. The species shown here grows to 3 ft. and has flowers about 1/2 in. across.

Widely distributed throughout our area, where it prefers moist woods and river banks, Blue Dogbane is cultivated as far north as Massachusetts.

Gray's manual recognizes *A. tabernaemontana* var. *gattingeri,* named in honor of its discoverer. It has narrower leaves than the typical variety and occurs in or adjacent to stream beds extending through cedar glades.

Flowering: April, May.

Angle-Pod *Gonobolus carolinensis* (Jacq.) Schul

Milkweed Family Asclepiadaceae

Angle-pod has a milky sap, simple entire opposite leaves, and chocolate-purple flowers in few-flowered umbels. Angle-pods (*Gonobolus* spp.) are vines with heart-shaped leaves and flowers that arise from the leaf nodes. This species has downy stems, 2 to 4 in.-wide leaves, and flowers (1/2 to 1 in. across) that are pollinated by flies.

This Angle-pod occurs in rich thickets (including the shrub zone of cedar glades) throughout the Central South.

Another local Angle-pod, *G. shortii,* has longer, narrower corolla lobes.

Flowering: June-August.

Periwinkle

Blue Dogbane

Angle-Pod

Butterfly-Weed

Asclepias tuberosa L.

Milkweed Family

Asclepiadaceae

One or more stems of this perennial plant (1 to 2 1/2 ft.) may emerge from a thickened root. Unlike other milkweeds, this species lacks a rubbery sap or latex. The numerous, somewhat hairy leaves are rather narrow and pointed. The small flowers vary from pale yellow-orange to brilliant orange-red. A common pollinator is the pink-edged Sulfur butterfly.

Butterfly-weed is found in dry, sunny sites as far south as Florida, north to New England, and west to Colorado. It is fairly frequent in eastern Tennessee and Kentucky but is less common in middle Tennessee.

Another common name, Pleurisy Root, refers to the use of the rhizomes (underground stems) by Indians and early settlers to treat respiratory diseases. It is now known to be toxic because of the presence of cardiac glycosides.

Flowering: June-September.

Antelope-Horn Milkweed

Asclepias viridis Walt.

Milkweed Family

[*Asclepiodora viridis* (Walt.) Gray]

Asclepiadaceae

Flowers of *Asclepias*, arranged in umbels, are intricate and distinctive. A typical milkweed flower has 5 reflexed petals above which are 5 erect hoods that form the corona (crown). In the Antelope-horn Milkweed, however, 1 in.-wide flowers include green petals that are spreading rather than reflexed; the crown is purplish.

Also called Spider-milkweed, this species is found in open dry habitats from Florida to Nebraska, and from eastern Kentucky to Ohio and West Virginia. It occurs frequently in deeper soils in glades in Tennessee, Alabama, and Georgia.

Another green-flowered milkweed also found in glades is the Green Milkweed, *A. viridiflora*, which has smaller flowers and reflexed petals.

Flowering: May, June.

White Milkweed

Asclepias variegata L.

Milkweed Family

Asclepiadaceae

This large milkweed (up to 3 ft. tall) is found in woods and thickets throughout most of the eastern U.S. The smooth broad leaves and flowers that form rounded umbels are typical of the genus.

The Whorled Milkweed, *A. verticillata*, also has white or greenish-white flowers in rounded umbels, but it has slender stems and whorled, linear leaves. It is an infrequent plant in glades and barrens in the region.

Flowering: May-July.

Butterfly-Weed

Antelope-Horn Milkweed

White Milkweed

Jacob's Ladder *Polemonium reptans* L.
Phlox Family Polemoniaceae

This is one of twenty *Polemonium* species in North America, but the only one in our area. Though the epithet *reptans* means "creeping," Jacob's Ladder is an upright plant 8 to 15 in. tall. Note the compound leaves with pointed leaflets and the small 1/2 in.-long flowers with white stamens.

Also called Greek Valerian, it grows in rich moist woods, especially along streams. It is found sporadically throughout the Central South but rarely in the Central Basin.

Flowering: April, May.

Blue Phlox *Phlox divaricata* L.
Phlox Family Polemoniaceae

Phloxes have flowers with long, narrow corolla tubes and 5 separate petals that extend at right angles. The flowers of Blue Phlox are nearly 1 in. across and vary from light blue to pale violet. The entire plant is usually less than 2 ft. tall.

This is a common and conspicuous wildflower in rich deciduous woods, especially on slopes. Though widespread throughout much of the eastern U.S., it is rare in the Central Basin.

Another phlox of moist woods, but found only along the eastern edge of our area, is Creeping Phlox, *P. stolonifera.* Unlike Blue Phlox, it has stamens that extend slightly beyond the corolla tube.

Flowering: April, May.

Glade Phlox *Phlox bifida* Beck ssp. *stellaria* (Gray) Wherry
Phlox Family Polemoniaceae

Asa Gray, the famous Harvard botanist, considered this to be a separate species (*P. stellaria* Gray), but it is now recognized only as a subspecies of *Phlox bifida.* Glade Phlox grows as a creeping mat, usually in the gravelly zone of open glades. The flowers are nearly 1/2 in. across and may be lavender, pink, light blue, or almost white.

Endemic to the Interior Low Plateau, this plant typically occurs in glades of the Central Basin.

The growth habit of Glade Phlox is similar to that of cultivated phloxes locally called Thrift. It can easily be grown in similar sunny situations—in sandy soil with good drainage.

Flowering: April, May.

Jacob's Ladder

Blue Phlox

Glade Phlox

Downy Phlox *Phlox pilosa* L.
Phlox Family Polemoniaceae

This slender perennial, 10 to 20 in. tall, has lance-shaped leaves. Note the bright pink wedge-shaped petals, each about 3/4 in. wide. "Downy" refers to the softly pubescent corolla tubes.

This phlox prefers dry, sunny locations such as roadsides and edges of woods. It occurs throughout the eastern U.S. but is not frequent in our area.

Flowering: April, May.

Purple Phacelia *Phacelia bipinnatifida* Michx.
Waterleaf Family Hydrophyllaceae

This upright biennial grows to a height of 2 ft. The mottled leaves are twice divided and coarsely-toothed; the bell-shaped flowers, each 1/2 in. across, have prominent stamens with white long-hairy filaments capped with rust-orange anthers.

It is rather frequent in rocky places within rich woods throughout our area.

One of several species called Scorpion-weed, *P. ranunculacea* is a much smaller plant with weak stems and larger flowers. A rare plant, it occurs sporadically in alluvial soils in our area but more commonly in the Mississippi floodplain, e.g., Reelfoot Lake.

Flowering: April, May.

Glade Scorpion-Weed *Phacelia dubia* (L.)
Waterleaf Family Trel. var. *interior* Fernald
Hydrophyllaceae

The famous Swedish botanist Carolus Linnaeus applied the name *dubia* because he was not sure whether this species was a *Phacelia* or *Polemonium*. A spreading annual, it has weak, branched stems supporting the white, light blue, or pink-tinged flowers with shallow cups (1/4 to 1/2 in. across).

Glade Scorpion-weed is found sporadically in woods and clearings throughout most of our area and north to Ohio and Pennsylvania. Variety *interior* is endemic to middle Tennessee where it forms masses in thin soils in cedar glades.

Fringed Phacelia, *P. purshii,* which also has bluish flowers, has delicately fringed petals. It occupies deeper glade soils.

Flowering: April, May.

Downy Phlox

Purple Phacelia

Glade Scorpion-Weed

Virginia Bluebells *Mertensia virginica* (L.) Pers.
Borage Family Boraginaceae

This beautiful plant stands 1 to 2 ft. tall; the leaves are smooth. The petals change from pink to bright blue as the flowers open. The arrangement of flowers is typical of the family: a one-sided, rolled-up coil gradually unfolds.

Virginia Bluebells often form solid stands along creek or river banks, or on islands. Found throughout much of the eastern U.S., they are infrequent in our area and probably do not occur in the Central Basin.

Flowering: March, April.

False Gromwell *Onosmodium molle* Michx.
Borage Family Boraginaceae

The word *false* in the common name distinguishes plants of this genus from those of *Lithospermum* (below), which are sometimes called Gromwell. Plants of this species are 1 to 2 ft. high, have softly hairy sessile leaves and conspicuous veins. The small tubular flowers often have a single style that extends beyond the corolla tube. The corolla may be greenish-white rather than yellowish as seen here.

Gattinger, in his *Flora of Tennessee*, notes False Gromwell as being "Abundant in the glades of M. Tenn." Often seen in pastures near glades where soils are thin, it is endemic to the Interior Low Plateau.

Flowering: April, May.

Hoary Puccoon *Lithospermum canescens* (Michx.) Lehm.
Borage Family Boraginaceae

This plant compensates for its small size (6 to 12 in. tall) by producing a dense inflorescence with numerous yellow-orange flowers. "Hoary" refers to the fine, soft hairs that cover the stem and leaves.

Hoary Puccoon grows especially well in calcareous soils; it is found in the glades of Tennessee and Alabama and also in the limestone regions of Kentucky.

Puccoon is an Indian name for a dye plant.

Flowering: April, May.

Virginia Bluebells

False Gromwell

Hoary Puccoon

Rose Verbena

Verbena canadensis (L.) Britt.

Verbena Family — Verbenaceae

Because this showy wildflower has weak stems, the plants are sprawling or only partially erect. The leaves are simple but deeply lobed and toothed like a fern frond. Each flower of the dense clusters has 5 petals that spread from the corolla tube, much like phloxes. Rose Verbena is pollinated by hawkmoths and various butterflies, including the strikingly beautiful swallowtails.

Rose Verbena grows on prairies and rocky roadsides from Florida to Texas and north to Illinois and Kansas. It is common in cedar glades but occurs less often elsewhere in disturbed sites in the Central South.

A weed also found along roadsides and in glades, Narrow-leaved Verbena, *V. simplex,* produces a 6 to 8 in. spike bearing tiny light blue flowers.

Flowering: April-June.

Wild Bergamot

Monarda fistulosa L.

Mint Family — Lamiaceae

Plants of the Mint Family, Lamiaceae or Labiaceae, have a distinctive combination of features. Their opposite leaves are aromatic; the corolla tube flairs to form 2 lips or labia, thus Labiatae.

Bergamots, *Monarda*, are erect herbs with toothed leaves and showy flower heads surrounded by bracts. Wild Bergamot grows 2 to 3 ft. tall. The flowers, each about 1 in. long, vary from white or pink to lavender or red.

This particular bergamot grows in dry open places, especially old fields and the edge of thickets. It occurs sporadically throughout our area, south to eastern Texas, and north to Canada.

The more attractive Bee-balm, *M. didyma,* with its bright red flowers, is usually found in the moist soils of uplands.

Fresh or dried leaves of bergamots are used for flavoring and for tea.

Flowering: June-August.

Glade Savory

Satureja glabella (Michx.) Briq.

Mint Family — Lamiaceae

This perennial is only 3 to 4 in. tall. It has smooth (*glabella* means "smooth") leaves and stems. Note the prominent calyx and the dark purple spots on the blue corolla.

Though it often grows at the edge of open cedar glades, Glade Savory is also found on moist limestone cliffs and on gravelly river banks throughout much of our area. It is endemic to the Interior Low Plateau.

The well-known herb basil is *S. vulgaris.*

Flowering: June-August.

Rose Verbena

Wild Bergamot

Glade Savory

False Dragonhead *Physostegia virginiana* (L.) Benth.
Mint Family Lamiaceae

This wiry, upright perennial can be 3 to 4 ft. tall. Each flower (1 in. long), varying in color from almost white to deep rose, has a spotted lower lip. The leaves (3 to 5 in. long) are lanceolate and have sharp teeth along their margins.

More likely to be seen in the eastern portions of our area, this wildflower occurs in barrens and moist waste places.

Flowering: August, September.

Downy Wood-Mint *Blephilia ciliata* (L.) Benth.
Mint Family Lamiaceae

Wood-mints, like bergamots (preceding page), are erect herbs with several dense whorls of bilabiate flowers at the upper leaf nodes. In the Downy Wood-mint, the flowers are light blue with purple spots; the bracts are purplish and showy. The leaves are sessile, or nearly so, narrowed to the base, and downy-white underneath. The stems have short, recurved hairs.

This plant occurs frequently in dry woods in limestone regions, especially in the Interior Low Plateau.

The Hairy Wood-mint, *B. hirsuta*, has stems with long spreading hairs and leaves with long petioles, rounded or cordate bases, and hair on both surfaces. Its niche is moist shady places.

Flowering: June.

Showy Skullcap *Scutellaria serrata* Andr.
Mint Family Lamiaceae

There are in our area about a dozen species of *Scutellaria*, otherwise known as skullcaps, so named for the small protuberance on the upper lip of the calyx suggesting a helmet or cap. Unlike most mints, skullcaps are not aromatic. The Showy Skullcap (1 to 2 ft. tall) has long (1 in.) flowers in a raceme. The specific epithet refers to saw-toothed margins of the ovate leaves.

The Showy Skullcap occurs sporadically in woods in the Appalachians and from Pennsylvania to Tennessee and North Carolina, including the Cumberland Plateau.

Flowering: May, June.

Small Skullcap *Scutellaria parvula* Michx.
Mint Family Lamiaceae

A diminutive (4 to 8 in.) wildflower, the Small Skullcap is about one-third the size of the preceding species. Note the spotted corolla and the leaves with scalloped margins. The stems have dense, short hairs and the leaves are sessile and ovate.

It is locally abundant in moist soils near the edges of cedar glades. Outside of glades, it may be found in other shallow limestone-derived soils.

Flowering: May, June.

False Dragonhead

Downy Wood-Mint

Showy Skullcap

Small Skullcap

Hairy Beard-Tongue *Penstemon hirsutus* (L.) Willd.
Figwort Family Scrophulariaceae

Members of the Figwort or Snapdragon Family have flowers in which the 5 (occasionally 4) petals are united into a bilabiate (2-lipped) corolla tube suggestive of a mint, but the minty fragrance is lacking.

Beard-tongues (*Penstemon* species) in our area are erect herbs with opposite, often sessile leaves. Flowers have 5 stamens, the longer, sterile one prominently bearded. In this species, the throat is nearly closed by the arched base of the lower lip. The stem is hairy. The flower color is invariably violet.

The Hairy Beard-tongue thrives in rocky places, usually with limestone-derived soils; it occurs mainly in the Bluegrass and Central Basin Regions, including cedar glades.

Flowering: May, June.

Slender-Flowered Beard-Tongue *Penstemon tenuiflorus* Pennell
Figwort Family Scrophulariaceae

A distinctive feature of this species is its lower lip bending upward to touch the upper one. The slender corolla tube is white or creamy yellow.

This beard-tongue is common in cedar glades in Kentucky, Tennessee, and Alabama. It also occurs in dry open woods and barrens in the Highland Rim.

Flowering: April, May.

Indian Paint-Brush *Castilleja coccinea* (L.) Spreng.
Figwort Family Scrophulariaceae

The scarlet-tipped, 3-lobed bracts are the showy parts of this 1 to 2 ft. plant. The yellowish flowers are all but hidden, except for the protruding pistils.

This species of *Castilleja,* the only one in our area, is found in meadows, prairies, and other open areas scattered throughout much of eastern North America. It is uncommon in the Central South, but ranges from New Hampshire to Florida and west to Oklahoma.

Wildflowers of the genus *Castilleja* and others of this family are parasitic on the roots of other plants.

Flowering: April, June.

Smooth False Foxglove *Gerardia laevigata* Raf.
Figwort Family [*Aureolaria laevigata* (Raf.) Raf.]
Scrophulariaceae

False foxgloves are wiry herbs known for their bright yellow bell-shaped flowers; the 5 equal lobes extend at right angles to the corolla tube. This species (3 to 5 ft. tall) has smooth lanceolate leaves with mostly entire margins. The lowest leaves sometimes have a few lobes and may be serrate.

Ranging from Pennsylvania and southern Ohio south to Georgia, Smooth False Foxglove is associated with dry or moist woods where it is parasitic on roots of oak trees. In our area, it is found mostly on the Cumberland Plateau.

The Downy False Foxglove, *G. virginica,* has hairy stems and leaves and leaf shapes that vary from deeply lobed throughout to lobed at the base and wavy margined or entire above.

Flowering: July, August.

Hairy Beard-Tongue

Slender-Flowered Beard-Tongue

Indian Paint-Brush

Smooth False Foxglove

Blue-Eyed Mary *Collinsia verna* Nutt.
Figwort Family Scrophulariaceae

Blue-eyed Mary is an annual that grows 2 ft. high. Its delicate (1/2 in.-long) bilabiate and bicolored flowers are unusual in that the middle lobe of the lower lip is folded lengthwise and encloses the 4 stamens. The sessile leaves are opposite and lanceolate.

This wildflower is locally abundant in rich woods. From New York and Wisconsin, its range extends into the Central South, including the Bluegrass Region, where it is fairly frequent. In Tennessee, it is protected by law.

Flowering: April, May.

Squaw Root *Conopholis americana* (L.) Wallr.
Broomrape Family Orobanchaceae

This parasitic plant, also called Cancer-root, occurs in clumps of thick, unbranched stems about 4 to 6 in. tall. The small bilabiate flowers, arranged on the axils of the scales, are white.

Squaw Root lives on roots of oaks and other deciduous trees. It is seen more often in the eastern part of our area.

Beech-drops, *Epifagus virginiana*, also of the Broomrape Family, has slender, branched stems 6 to 18 in. tall. Although found only in association with beech trees, it is probably more common than Squaw Root.

Flowering: May, June.

Hairy Ruellia *Ruellia caroliniensis* (Walt.) Steud.
Acanthus Family Acanthaceae

In our area, wildflowers of the genus *Ruellia* are 1 to 3 ft. tall with opposite leaves and blue-violet trumpetlike flowers in their axils. From each slender corolla tube extend 5 flaring lobes. At the base of each flower is a pair of small leaves. Hairy Ruellia has hairy stems and leaves. The leaves are sessile as are the flowers, which have long corolla tubes.

This widely distributed wildflower, found in woods and thickets throughout our area, is less common in the Bluegrass Region and the Central Basin.

Smooth Ruellia, *R. humilis*, is common in cedar glades but is found in other dry limestone habitats. It is a smooth plant with stalked flowers and leaves. Its stems are often reclining.

Flowering: May-October.

Large Houstonia *Houstonia purpurea* L. var. *calycosa* Gray
Madder Family [*H. lanceolata* (Poir.) Britt]
Rubiaceae

Most plants of this genus are delicate wildflowers called bluets. This species (6 to 12 in. tall) has pinkish or violet flowers clustered above the uppermost pair of sessile, oblong-lanceolate leaves.

Large Houstonia is found throughout much of the Central South. In cedar glades, this variety of the species occurs in deeper soils or along stream banks.

Bluets in our region include *H. serpyllifolia*, *H. caerulea*, *H. nigricans*, and *H. patens*, the last three of which sometimes occur in cedar glades.

Flowering: May, June.

Blue-Eyed Mary

Squaw Root

Hairy Ruellia

Large Houstonia

Cardinal Flower *Lobelia cardinalis* L.
Bluebell Family Campanulaceae

Lobelias are tall wandlike herbs with alternate, lanceolate leaves. All species of *Lobelia* in our area have blue flowers except for the spectacular Cardinal Flower. This 2 to 4 ft.-tall plant has 1 to 1 1/2 in.-long flowers. Extending from the tubular corolla of each flower are 2 upper narrow lobes and 3 wider lobes below.

Cardinal Flower ranges from southeastern Canada to the southern part of our area. It is invariably found near water.

Flowering: July-September.

Gattinger's Lobelia *Lobelia appendiculata* A. DeCondolle var. *gattingeri* (Gray) McVaugh [*L. gattingeri* Gray]
Bluebell Family Campanulaceae

Named in honor of a pioneer botanist of Tennessee, Gattinger's Lobelia is rather small (6 to 12 in.) with mostly unbranched stems. The sessile leaves with clasping bases are arranged along the lower part of the stem. The flowers are 1/2 in. long or less.

The species *L. appendiculata* is widely distributed in prairies and open pine woods from the southern Midwest eastward through our area. This variety is an endemic known only from cedar glades in the Central Basin.

The Pale Spiked Lobelia, *L. spicata,* is a somewhat larger plant (1 to 2 ft.) with pale blue flowers. It is found in cedar glades and sporadically in oak or cedar woods in the Central South.

Because of the presence of certain alkaloid substances, several lobelias have been used medicinally or as narcotics.

Flowering: May, June.

Tall Bellflower *Campanula americana* L.
Bluebell Family Campanulaceae

Species of *Campanula* have regular flowers. In this species, the petals are not bell-shaped but are widely spreading and about 1 in. across. Also note the long, curved style characteristic of this tall (2 to 6 ft.) annual.

It is fairly common in moist soils throughout the Central South and north to Ontario, west to Oklahoma, and south to Florida.

Flowering: May-July.

Wild Teasel *Dipsacus sylvestris* Huds.
Teasel Family Dipsacaceae

Flowering heads of teasel look like those of composites (Asteraceae). One distinction is the lack of fusion of the stamens in the tiny florets of teasel. Numerous small, spiny bracts project beyond the white-to-lavender florets.

Wild Teasel, a European plant, naturalized throughout much of eastern North America, grows as a weed on roadsides and old fields.

The dried heads of Fuller's Teasel, var. *fullonum* of this species, are useful commercially in giving a desirable finish to textiles.

Flowering: June-October.

Cardinal Flower

Gattinger's Lobelia

Tall Bellflower

Wild Teasel

Yarrow *Achillea millefolium* L.

Daisy Family Asteraceae

The Asteraceae or Compositae, as it is traditionally called, is the largest plant family in North America. Composites are variable, but the arrangement of many tiny flowers (florets) on a common receptacle to form a head makes them recognizable. Each floret typically has a floral formula (Introduction, Part 3) of 5-5-5-1. A unique feature of composite flowers is the fusion of the anthers to form a cylinder around the style; above the style is a 2-lobed stigma.

In Yarrow, the florets form small heads (1/2 in. across) arranged into a flat-topped cluster. The plants (1 to 3 ft.) have lacy, fernlike leaves.

This European weed is widely scattered throughout North America in fields, waste places, and other habitats.

American Indians and, later, Europeans used Yarrow for a host of medicinal purposes. Modern studies have confirmed its efficacy as an astringent and anti-inflammatory agent. It is also cultivated as an ornamental.

Flowering: May-August.

Wild Quinine *Parthenium integrifolium* L.

Daisy Family Asteraceae

Each of the 5-ray flowers surrounding the head is about 1/2 in. long. The leaves, alternately attached to the stem, are sessile and lanceolate and have a toothed margin.

Wild Quinine occupies open woods in our area and Midwestern prairies.

Leaves of this plant were used by the Catawba Indians in the Carolinas to treat burns. The common name apparently refers to its bitter taste.

Flowering: May-September.

Plantain-Leaved Pussytoes *Antennaria plantaginifolia* L. Hook.

Daisy Family Asteraceae

Pussytoes are small perennials, usually with several small, soft, downy heads clustered together. The heads on a plant possess either male or female florets. In this species, the male plants are 1 to 6 in. tall whereas the female ones are 2 to 10 in. Each leaf has 3 main veins.

This wildflower occurs often in sizable populations, in fields, roadsides, and open woods throughout the eastern U.S. Preferring a somewhat acidic soil, it is uncommon in the Central Basin.

Field Pussytoes, *A. neglecta,* is similar except for its smaller, 1-veined leaves. Solitary Pussytoes, *A. solitaria,* has only 1 head on a nearly naked stem.

Flowering: April, May.

Daisy Fleabane *Erigeron philadelphicus* L.

Daisy Family Asteraceae

Fleabanes are sometimes weedy herbs, often with downy or hairy leaves and stems. Daisy Fleabane (about 2 ft. tall) has several heads per stem, each 3/4 in. wide, with numerous white-to-pinkish rays surrounding the yellow disc. The leaves are generally toothed; the upper ones have clasping bases.

It is common especially in moist fields and ditches throughout most of our area and the eastern U.S.

Flowering: May, June.

Yarrow

Wild Quinine

Plaintain-Leaved Pussytoes

Daisy Fleabane

Thin-Leaved Sunflower *Helianthus decapetalus* L.
Daisy Family Asteraceae

In addition to the cultivated Common Sunflower, *H. annuus,* there are about 24 sunflowers in the Central South. These native species are generally tall, coarse perennials with simple, rough-textured leaves. The Thin-leaved Sunflower has a rather smooth stem, 2 to 5 ft. tall, and lanceolate leaves with blade bases tapering to form winged petioles. The heads are about 2 1/2 in. across and usually have 8 to 12 rays.

Fairly common in the Cumberland Plateau, the Thin-leaved Sunflower occurs occasionally and sporadically in other portions of our area.

The Woodland Sunflower, *H. divaricatus,* has rougher, nearly sessile leaves (hairy below) that lack the winged petioles; its heads often have more (8 to 15) rays. It is probably more widespread within our area and occupies drier habitats like cedar glades.

Flowering: July-September.

Autumn Sneezeweed *Helenium autumnale* L.
Daisy Family Asteraceae

Note the heads, approximately 1 in. across, with globular yellow-gray discs and the slightly reflexed, wedge-shaped rays. This 2 to 5 in.-tall plant has winged stems and toothed, lanceolate leaves.

Preferring moist soil, this showy wildflower is found in swamps, wet meadows, and along the edge of streams. It occurs frequently throughout our area.

Purple-headed Sneezeweed, *H. flexuosum,* is similar but has purplish-brown discs. Bitterweed, *H. amarum,* is shorter (1 ft.) and has heads similar to *H. autumnale,* but very narrow, grasslike leaves. Bitterweed is common in disturbed soil, especially along roadsides and in overgrazed pastures. It is usually avoided by animals, but may cause cows eating it to give milk with a bitter taste.

Flowering: August-October.

Lance-Leaved Gumweed *Grindelia lanceolata* Nutt.
Daisy Family Asteraceae

Gumweeds are tough perennials with several branches bearing yellow daisylike flower heads, 1 in. across. Gumweed describes the involucre whorls of overlapping gummy green bracts at the base of each head. Various stages in the floral development of *G. lanceolata* are seen in this photo. Note the linear-lanceolate leaves.

In our area, Lance-leaved Gumweed is found principally in cedar glades and rocky roadsides in Tennessee and Alabama. Its range extends to Texas and Missouri.

Another gumweed, *G. squarrosa,* a more widespread species in mid-continental prairies, is occasionally seen in our area. It has wider leaves and more (24 to 40 vs. 16 to 20) rays per head.

Flowering: August-October.

Thin-Leaved Sunflower

Autumn Sneezeweed

Lance-Leaved Gumweed

Tickseed-Sunflower *Bidens aristosa* (Michx.) Britt.
Daisy Family Asteraceae

The showiest member of the *Bidens* group, Tickseed-sunflower is 2 to 4 ft. tall and has large compound leaves, each divided into 5 narrow, toothed leaflets. The flower heads are 1 to 2 in. wide and have 6 to 10 rays each. In the illustration, the insects at the right are soldier beetles.

This wildflower is often seen in large populations in open wet fields and roadsides. It is widespread throughout the Central South, north to New England, and west to Texas. "Sticktight" is a name appropriately applied to the species. The numerous double-awned, barbed achenes (small, dry, one-seeded fruits) are scattered as they adhere to animal fur or to clothing.

In cedar glades, you may find Spanish-needles, *B. bipinnata*. Its heads have much smaller yellow rays and its leaves wider, incised leaflets. Along with a dozen or so other less showy species of *Bidens*, it is found throughout our area.

Flowering: August-October.

Black-Eyed Susan *Rudbeckia hirta* L.
Daisy Family Asteraceae

Known by many simply as coneflowers, *Rudbeckia* species are characterized by brown dome-shaped discs and long yellow rays. Black-eyed Susan is a bristly 1 to 3 ft.-tall plant with one to several heads. The heads, measuring 2 to 3 in. wide, have 10 to 30 rays.

Black-eyed Susans brighten roadsides, old fields, and open woods. They are quite prevalent in most of our area, except for the Bluegrass Region.

Also common, especially in the Bluegrass Region and Central Basin, is the Thin-leaved Coneflower, *R. triloba*. It has smaller, more numerous heads on branched stems. The lower leaves on the stem are 3-lobed.

Flowering: July-September.

Broom-Weed *Xanthocephalum dranunculoides* (DC) Shinners
Daisy Family [*Guitierrezia dranunculoides* (DC) Blake]
Asteraceae

The thin, brittle, broomlike stems of this resinous, bushlike annual account for its common name. The small but numerous flower heads have 5 to 10 rays each and measure 1/4 to 1/2 in. across. The stems are 2 to 3 ft. tall.

In 1973, Dr. Robert Kral of Vanderbilt University pointed out that this plant has only recently become common in cedar glades (especially in Rutherford, Davidson, and Wilson Counties of the Central Basin). Elsewhere, it is known from northern Alabama and in calcareous habitats just west of the Central South.

Flowering: July-October.

Tickseed-Sunflower

Black-Eyed Susan

Broom-Weed

Tall Goldenrod

Solidago altissima L.

Daisy Family

Asteraceae

Of the approximately 90 goldenrods in North America, almost half occur in our area. The Tall Goldenrod's grayish stem may reach 6 ft. in height. The heads are clustered on arching side branches that may be 1 ft. long. The lanceolate leaves are rough on the upper sides and hairy underneath.

This goldenrod is common throughout most of its range, which covers nearly all of the eastern U.S. and adjacent Canada and extends to Wyoming.

Contrary to popular belief, goldenrods seldom cause hay fever. The pollen of ragweeds, *Ambrosia* species, produced in much larger amounts during the same time that goldenrods are in flower, is the usual cause.

Flowering: August-October.

Short's Goldenrod

Solidago shortii T. and G.

Daisy Family

Asteraceae

In comparison with the Tall Goldenrod, Short's is only 2 to 4 ft. tall and lacks the grayish stems. The rare plant is endemic to cedar glades and rocky slopes in central Kentucky.

Of some historical significance is Gattinger's Goldenrod, *S. gattingeri.* Named by the famous botanist Alvan W. Chapman, a 19th-century contemporary of the Nashville-based Gattinger, it is known primarily from Missouri. However, there was also a record of its occurrence at Lavergne (Rutherford County), Tennessee.

Flowering: July-September.

Prairie Coneflower

Ratibida pinnata (Vent.) Barnh.

Daisy Family

Asteraceae

Also known as Gray-headed Coneflower, this 3 to 4 ft.-tall wildflower is distinctive. As seen here, the cones are elongated; with age their color changes from greenish to dark gray or brown. The 2 in.-long rays are extremely reflexed; their ends often touch the supporting stalk. As suggested by *pinnata,* the leaves are deeply dissected.

The Prairie Coneflower occupies open grassy areas, roadsides, and limestone glades in Tennessee and Georgia. From our area, its range extends north to Ontario and Nebraska and west to Oklahoma.

The Green-headed Coneflower, *Rudbeckia laciniata,* is similar in overall appearance. Its discs, though, are more dome-shaped, its rays less reflexed, and its leaf lobes wider. It typically grows in moist places.

Flowering: May-August.

Tall Goldenrod

Short's Goldenrod

Prairie Coneflower

Small's Ragwort

Senecio smallii Britt.

Daisy Family

Asteraceae

Small's Ragwort is a perennial with a 1 1/2 to 2 ft. stem that is woolly toward the bottom. The basal leaves are variable but generally lanceolate, toothed and/or dissected. The flower heads are about 1 in. across.

Named in honor of botanist John K. Small, this wildflower is common in open woods and clearings from Florida to New Jersey, and to Tennessee and Indiana. In our area it is more common in the eastern portion, but is widespread in the deeper soils in open cedar glades.

Two other ragworts are also common: Golden Ragwort, *S. aureus,* has nearly round, cordate basal leaves; and Butterweed, *S. glabellus,* is a taller plant (up to 3 ft.) with thick, hollow stems and much larger, dissected leaves continuing up the stem. The latter is restricted to wet sites.

Several of the more than 2,000 species of *Senecio,* especially *S. aureus,* have been used by herbalists for a variety of medicinal purposes: as a "female regulator," as treatment for kidney stones, and as a diuretic. However, it is now known that these plants contain alkaloids toxic to the liver.

Flowering: April, May.

Cut-Leaved Prairie-Dock

Silphium terebinthinaceum

Daisy Family

[Jacq. var. *pinnatifidum* (Ell.) Gray]

Asteraceae

Because of their resinous sap, the large, coarse plants of the genus *Silphium* are often called rosinweeds. This species (3 to 4 ft. tall) has large basal leaves, up to 1 1/2 ft. long. In this variety they are deeply dissected.

Rosinweeds are prairie plants. This particular plant, photographed at May Prairie near Manchester, Tennessee, is widely distributed in the eastern U.S. Within the Central South, this variety occurs in barrens and rarely in glades.

The Whorled Rosinweed, *S. trifoliatum,* the more typical rosinweed in cedar glades, also occurs in open woods in most of our area. The rough, petioled leaves of this plant are lanceolate and arranged 3 to 5 per node.

Flowering: July-September.

Bushy Aster

Aster dumosus L.

Daisy Family

Asteraceae

Bushy Aster is a spreading plant 1 to 3 ft. tall. In addition to linear or narrowly oblong leaves attached to the main stem, there are many tiny bractlike leaves on the branches.

This common plant is found in old fields and other open habitats throughout much of the eastern U.S.

Of the 30 or so additional asters in our area, *A. pilosus* is probably the most similar and most common. However, it has a less compact growth habit and white (or lighter blue) ray flowers. It is common in both old fields and cedar glades.

Flowering: August-October.

Small's Ragwort

Cut-Leaved Prairie-Dock

Bushy Aster

Joe Pye-Weed *Eupatorium fistulosum* Barrett
Daisy Family Asteraceae

Several species of *Eupatorium* share the name Joe Pye-weed. They are sometimes 6 ft. or more tall, topped with massive clusters of small pink-purple flower heads. In this species, the inflorescence is rounded and composed of heads with 5 to 8 flowers per head. The stems are purplish and hollow; the leaves are in whorls of 5 to 7.

This impressive wildflower inhabits stream banks and other moist places throughout most of the eastern U.S.

Similar species in our area, but with solid stems, include Sweet Joe Pye-weed, *E. purpureum,* which has green stems and fewer leaves per node, and Spotted Joe Pye-weed, *E. maculatum,* with its dark purple or purple-spotted stems and nearly flat-topped flower clusters.

Joe Pye was an Indian medicine man who became famous in Colonial New England for his use of *E. purpureum* in the treatment of typhus.

Flowering: July-September.

Nodding Thistle *Carduus nutans* L.
Daisy Family Asteraceae

Also called Musk Thistle, the Nodding Thistle is the most conspicuous thistle in our region. Note the large reflexed bracts surrounding the base of each nodding head. The stems and leaves are spiny.

Since its introduction, this European biennial has been spreading from the northeastern U.S. westward and southward. It is now common in disturbed soil throughout the Central South.

Most other thistles in our area are *Cirsium* species. Included is Bull Thistle, *C. vulgare,* which has similar but not-nodding heads with clusters of upturned bracts.

This aggressive plant is often a serious agricultural problem. However, the fluffy down that disperses the seeds is used by the Yellow Warbler to line its nest.

Flowering: June-August.

New York Ironweed *Vernonia noveboracensis* (L.) Michx.
Daisy Family Asteraceae

Ironweeds resemble Joe Pye-weeds (above), but have flower heads tending toward a deeper shade of purple or violet and alternate (rather than whorled) leaves. The New York Ironweed is tall (3 to 6 ft.) and has slender, sessile leaves. Note the hairlike tips on the bracts surrounding each flower head.

It is found in the Northeast, but its range extends as far south as Georgia. In our area, it is more common in the Highland Rim than elsewhere. It grows in low, moist places.

Probably more common in our area is the Tall Ironweed, *V. altissima.* It may become taller but, more important diagnostically, has leaves that are downy underneath and bracts lacking the tips described above.

Flowering: July-October.

Joe Pye-Weed

Nodding Thistle

New York Ironweed

Rough Blazing-Star *Liatris aspera* Michx.
Daisy Family Asteraceae

The showy pink or purplish flowers of blazing-stars on tall spikes punctuate mid-continental North American prairies. Some *Liatris* species extend into grassy glades and prairies in the Central South. Note the closely spaced heads of *L. aspera,* each with 25 to 40 florets and sessile or nearly so. The rows of round-tipped green bracts distinguish this species from others.

Rough Blazing-star, which occupies rocky open areas including cedar glades, occurs sporadically within the Central South.

Also infrequent in these habitats is Scaly Blazing-star, *L. squarrosa,* which has fewer heads and spreading bracts with pointed tips.

Flowering: August-October.

Pale Purple Coneflower *Echinacea pallida* Nutt.
Daisy Family Asteraceae

Flower heads combining rust-colored conical discs and prominent lavender-to-purple rays characterize the North American genus *Echinacea.* As with *Liatris* (above), these plants are commonly associated with prairies. *Echinacea pallida* as seen here has spreading or drooping rays (2 to 3 in. long). The leaves are lanceolate-linear, parallel-veined, and lack teeth.

It is seen occasionally in rocky glades and barrens in the Central Basin and in the Highland Rim of Kentucky.

Purple Coneflower, *E. purpurea,* occurs infrequently in moist soil in open woods in the Highland Rim and Central Basin. It usually has darker rays and toothed, lanceolate to broadly ovate leaves.

Flowering: June, July.

Tennessee Coneflower *Echinacea tennesseensis* (Beadle) Small
Daisy Family Asteraceae

This coneflower resembles *E. pallida* (above) rather closely, but has pinkish-purple rays that are usually spreading or even slightly upturned rather than reflexed.

It lacks an effective means of dispersing its seeds from one open glade to another. Thought in the 1960s to be extinct, it is now recognized as endemic to the cedar glades in three Central Basin counties.

Tennessee Coneflower, listed as a federally endangered species, is protected by state and federal laws prohibiting taking without the landowner's permission or sale without a state license or federal permit.

American Indians used the rhizomes of the *Echinacea* species to treat animal bites and toothaches. Physicians of the 19th and early 20th centuries considered it useful as an antiseptic and a remedy for infections. *Echinacea* is currently being reevaluated, especially in Europe, for its medicinal value.

Flowering: May-October.

Rough Blazing-Star

Pale Purple Coneflower

Tennessee Coneflower

Bibliography

Popular Illustrated Books

Dean, Blanche E., Amy Mason, and Joab L. Thomas. 1973. *Wildflowers of Alabama and adjoining states.* University, Ala: University of Alabama Press.

Dobelis, Inge N., ed. 1986. *Magic and medicine of plants.* Pleasantville, N.Y.: Reader's Digest Association, Inc.

Duncan, Wilbur H., and Leonard E. Foote. 1975. *Wildflowers of the southeastern United States.* Athens: University of Georgia Press.

Foster, Steven, and James A. Duke. 1990. *Field guide to medicinal plants, eastern and central North America.* Boston: Houghton Mifflin.

Kricher, John C., and Gordon Morrison. 1988. *A field guide to eastern forests.* Boston: Houghton Mifflin.

Mohlenbrock, Robert H. 1987. *Wildflowers, a quick identification guide to the wildflowers of North America.* New York: Collier Books.

Peterson, Lee. 1978. *A field guide to edible wild plants.* Boston: Houghton Mifflin.

Peterson, Roger T., and Margaret McKenny. 1968. *A field guide to wildflowers of northeastern and northcentral North America.* Boston: Houghton Mifflin.

Smith, Arlo I. 1979. *A guide to wildflowers of the mid-south.* Memphis: Memphis State University Press.

Sutton, Ann, and Myron Sutton. 1985. *Eastern forests.* New York: Chanticleer Press.

Wharton, Mary E., and Roger W. Barbour. 1971. *A guide to the wildflowers and ferns of Kentucky.* Lexington: University Press of Kentucky.

Williams, John G., and Andrew E. Williams. 1983. *Field guide to orchids of North America.* New York: Universe Books.

Scientific Publications

Baskin, Jerry, and Carol C. Baskin. 1985. Life cycle ecology of annual plant species of cedar glades of southeastern United States. In *The population structure of vegetation,* ed. J. White. Dordrecht: Dr. W. Junk Pub.

Baskin, Jerry, and Carol C. Baskin. 1988. Endemism in rock outcrop plant communities of unglaciated eastern United States: An evaluation of the roles of the edaphic, genetic and light factors. *J. Biogeogr.* 15:829-40.

Baskin, Jerry M., Elsie Quarterman, and Carol Caudle. 1968. Preliminary check-list of the herbaceous vascular plants of cedar glades. *J. Tenn. Acad. Sci.* 43:65-71.

Braun, E. Lucy. 1950. *Deciduous forests of eastern North America.* New York: Blakiston.

Fernald, M. L. 1950. *Gray's manual of botany.* 8th ed. New York: American Book.

Gattinger, Augustin. 1901. *Flora of Tennessee and philosophy of botany.* Nashville: Gospel Advocate Pub. Co.

Gibson, D. 1961. Life-forms of Kentucky flowering plants. *Amer. Midl. Nat.* 66:1-60.
Hemmerly, Thomas E., and Elsie Quarterman. 1978. Optimum conditions for the germination of seeds of cedar glade plants: a review. *J. Tenn. Acad. Sci.* 53:7-11.
Kuchler, A. W. 1964. Potential natural vegetation of the conterminous United States. *Amer. Geogr. Soc. Spec. Publ. No. 36.*
Quarterman, Elsie. 1950. Major plant communities of Tennessee cedar glades. *Ecol.* 31:234-54.
———. 1973. Allelopathy in cedar glade plant communities. *J. Tenn. Acad. Sci.* 43:147-50.
Quarterman, Elsie, and Richard L. Powell. 1978. *Potential ecological/geological natural landmarks on the Interior Low Plateaus.* Denver: Division of Natural Landmarks.
Tennessee Dept. of Finance and Administration. 1966. *Tennessee, its resources and economy, vol. 2: Tennessee natural resources.* Publ. No. 33lb. Nashville.

Symposia

Baskin, Jerry M., Carol C. Baskin, and Ronald L. Jones, eds. 1987. Symposium: the vegetation and flora of Kentucky. Richmond: Kentucky Native Plant Society.
Chester, E. W., ed. 1989. The vegetation and flora of Tennessee. Proceedings of a symposium sponsored by the Austin Peay State University Center for Field Biology. *J. Tenn. Acad. Sci.* 64, no. 3.
Somers, Paul, ed. 1986. Symposium: biota, ecology, and ecological history of cedar glades. *ASB Bull. No. 33.*

Index

Acanthaceae, 96
Acanthus Family, 96
Achillea millefolium, 100
Agavaceae, 32
Agave, 32
Agave virginica, 32
Alliaria officinalis, 54
Allium, 28
Allium canadense, 26
Allium cepa, 26
Allium cernuum, 26
Allium stellatum, 26
Allium vineale, 26
Amaryllidaceae, 32
Ambrosia, 106
Amsonia tabernaemontana, 80
Andropogon gerardi, 20
Andropogon virginicus, 20
Anemonella thalictroides, 50
Angiosperms, 13, 16
Angle-pod, 80
Antennaria neglecta, 100
Antennaria plantaginifolia, 100
Antennaria solitaria, 100
Apiaceae, 76
Apocynaceae, 80
Aquilegia canadensis, 50
Araceae, 24
Araliaceae, 76
Arenaria patula, 46
Arisaema dracontium, 24
Arisaema triphyllum, 24
Aristida longespica, 20
Aristolochiaceae, 44
Arum Family, 24
Asarum arifolium, 44
Asarum canadense, 44
Asclepiadaceae, 80
Asclepias tuberosa, 82
Asclepias variegata, 82
Asclepias verticillata, 82
Asclepias viridiflora, 82
Asclepias viridis, 82
Asclepiodora viridis, 82
Ash, Blue, 11
Aster, Bushy, 108
Aster dumosus, 108
Aster pilosus, 108
Asteraceae, 40-46
Astragalus bibullatus, 62
Astragalus canadensis, 62
Astragalus tennesseensis, 62
Aureolaria laevigata, 94

Balsaminaceae, 68
Baptisia australis, 62
Baptisia leucantha, 62
Baptisia tinctoria, 62
Barberry Family, 52
Baskin, Carol, 9, 56
Baskin, Jerry, 9, 56
Beard-tongue, Hairy, 94
Beard-tongue,
  Slender-flowered, 94
Bee-balm, 90
Beech-drops, 96
Belamcanda chinensis, 34
Bellflower, Tall, 98
Berberidaceae, 52
Bergamot, Wild, 90
Bidens aristosa, 104
Bidens bipinnata, 104
Birthwort Family, 44
Bitterweed, 102
Blackberry-lily, 34
Bladderpod, Duck River, 54
Bladderpod, Stone's River, 54
Blazing-star, Rough, 112
Blazing-star, Scaly, 112
Blephilia ciliata, 92
Blephilia hirsuta, 92
Bluebell Family, 98
Bluebells, Virginia, 88
Blue-eyed Grass, 34
Blue-eyed Grass, White, 34
Blue-eyed Mary, 96
Bluegrass Basin, 2, 4
Blue-star, 80
Bluestem, Big, 20
Bluestem, Little, 20
Bluets, 96
Borage Family, 88
Boraginaceae, 88
Brassicaceae, 54, 56
Braun, E. Lucy, 5, 30
Breadroot, Nashville, 60
Brier, Sensitive, 60
Broomrape Family, 96
Broom-sedge, 20
Broom-weed, 104
Buckwheat Family, 42
Buttercup Family, 46-50
Buttercup, Hooked, 48
Buttercup, Kidney-leaf, 46
Buttercup, Tall, 46
Butterfly-weed, 82
Butterweed, 108

Cactaceae, 72
Cactus Family, 72
Calopogon pulchellus, 36
Calopogon tuberosus, 36
Campanula americana, 98
Campanulaceae, 98
Cancer-root, 96
Cardinal Flower, 98
Carduus nutans, 110
Carduus vulgare, 110
Carex frankii, 22
Carya, 5, 11
Caryophyllaceae, 46
Castilleja coccinea, 94
Catawba Indians, 100
Catchfly, Round-leaved, 46
Cattail, Common, 20
Cattail Family, 18, 20
Cattail, Narrow-leaved, 20
Cedar glades, ix, 7-9, 11-12
Celtis laevigata, 11
Central Basin, 2, 3, 7
Central Highlands, 3, 4
Chapman, Alvin, 106
Chufa, 22
Cichorium intybus, 1
Cinquefoil, Dwarf, 58
Cinquefoil, Rough-fruited, 58
Cinquefoil, Sulphur, 58
Cirsium vulgare, 110
Claytonia caroliniana, 44
Claytonia virginica, 44
Clematis glaucophylla, 50
Clematis pitheri, 50
Clematis reticulata, 50
Clematis versicolor, 50
Clematis viorna, 50
Coastal Plain, 2, 4
Collinsia verna, 96
Columbine, Wild, 50
Columbo, American, 78
Commelina communis, 24
Commelina erecta, 24
Commelinaceae, 24
Common chicory, 1
Compositae, 100
Composite Family, 16, 100
Coneflower, Gray-headed, 106

Coneflower, Green-headed, 106
Coneflower, Pale Purple, 112
Coneflower, Prairie, 106
Coneflower, Purple, 112
Coneflower, Tennessee, 9, 12, 112
Coneflower, Thin-leaved, 104
Conium maculatum, 76
Conopholis americana, 96
Coralberry, 11
Coral-root, Autumn, 36
Coral-root, Spring, 36
Corallorhiza odontorhiza, 36
Corallorhiza wisteriana, 36
Cranesbill, Carolina, 64
Crassulaceae, 58
Croton capitatus, 66
Croton monanthogynus, 66
Croton, Woolly, 66
Cruciferae, 54
Cumberland Plateau, 2, 4-5
Cyperaceae, 22
Cyperus esculenta, 22
Cyperus inflexus, 22
Cyperus papyrus, 22
Cyperus strigosus, 22
Cypripedium acaule, 38
Cypripedium calceolus, 38
Cypripedium reginae, 38

Daffodil Family, 32
Daisy Family, 100-12
Dalea gattingeri, 62
Dalea purpureum, 62
Daucus carota, 1
Dayflower, Asiatic, 24
Dayflower, Slender, 24
Delphinium carolinianum, 48
Delphinium tricorne, 48
Delphinium virescens, 48
Dentaria diphylla, 56
Dentaria laciniata, 56
Dentaria multifida, 56
Desmanthus illinoensis, 60
Dicentra canadensis, 52
Dicentra cucullaria, 52
Dicots, 16, 41
Dipsacaceae, 98
Dipsacus sylvestris, 98
Dodecatheon meadia, 76
Dogbane, Blue, 80
Dogbane Family, 80
Dragon, Green, 24
Dragonhead, False, 92
Dutchman's Breeches, 52

Echinacea pallida, 112
*Echinacea purpurea*112
Echinacea tennesseensis, 112
Eggleston, Willard W., 70
Eleocharis compressa, 22
Elm, Winged, 11
Epifagus virginiana, 96
Erigeron philadephicus, 100
Erythronium albidum, 28
Erythronium americanum, 28
Eupatorium fistulosum, 110
Eupatorium maculatum, 110
Eupatorium purpureum, 110
Euphorbia corollata, 66
Euphorbiaceae, 66
Evening-primrose, Common, 74
Evening-primrose Family, 74
Evening-primrose, Missouri, 74
Evening-primrose, Showy, 74
Evening-primrose, Three-lobed, 74
Evening-primrose, White, 74

Fabaceae, 60, 62
Fairy Spuds, 44
False Aloe, 32
False Foxglove, Downy, 94
False Foxglove, Smooth, 94
False-garlic, 28
Fame Flower, Limestone, 11, 44
Figwort Family, 94, 96
Fire Pink, 46
Fleabane, Daisy, 100
Flora of Tennessee, 9, 88
Flowers, 16
Foamflower, 58
Forbs, 12
Forestiera ligustrina, 11
Fraxinus quadrangulata, 11

Garlic, Field, 26
Garlic, Wild, 26
Garlic-mustard, 54
Gattinger, Augustin, 9
Gentianaceae, 78
Gentian Family, 78
Geraniaceae, 64
Geranium carolinianum, 64
Geranium Family, 64
Geranium maculatum, 64
Geranium, Wild, 64
Gerardia laevigata, 94
Gerardia virginica, 94
Ginseng, 76
Ginseng, Dwarf, 76
Ginseng Family, 76
Glade-cress, 56
Glade Moss, 1, 11
Goat's-Rue, 60
Goldenrod, Gattinger's, 106
Goldenrod, Short's, 106
Goldenrod, Tall, 106
Golden-seal, 48
Gonobolus carolinensis, 80
Gonobolus shortii, 80
Grass Family, 20
Grass,
Long-spiked Three-awn, 20
Gray, Asa, 18
Greek Valerian, 84
Grindelia lanceolata, 102
Grindelia squarrosa, 102
Gromwell, False, 88
Ground Plum, Guthrie's, 62
Guitierrezia dranunculoides, 104
Gum, Sweet, 12
Gumweed, 102
Gumweed, Lance-leaved, 102
Guttiferae, 68

Habenaria ciliaris, 36
Hackberry, 11
Helenium amarum, 102
Helenium autumnale, 102
Helenium flexuosum, 102
Helianthus annuus, 102
Helianthus decapetalus, 102
Helianthus divaricatus, 102
Hemlock, Poison, 76
Hepatica acutiloba, 48
Hepatica, Sharp-lobed, 48
Hibiscus militaris, 68
Hibiscus moscheutos, 68
Hickories, 11
Highland Rim, 2, 3-4, 6, 12
Houstonia caerulea, 96
Houstonia lanceolata, 96
Houstonia, Large, 96
Houstonia nigricans, 96
Houstonia patens, 96
Houstonia purpurea, 96
Houstonia serpyllifolia, 96
Hydrastis canadensis, 48
Hydrophyllaceae, 86
Hymenocallis caroliniana, 32
Hymenocallis occidentalis, 32
Hypericum dolabriforme, 68
Hypericum frondosum, 11
Hypericum perforatum, 68
Hypericum sphaerocarpum, 68
Hypericum turgidum, 68
Hypoxis hirsuta, 32

Impatiens capensis, 68
Impatiens pallida, 68
Indian-pink, 78
Indian Turnip, 24
Indigo, Blue False, 62
Indigo, White False, 62
Indigo, Wild, 62
Indigofera tinctoria, 62
Interior Low Plateau, 4
Iridaceae, 34
Iris, Bearded, 34
Iris, Crested Dwarf, 34

Iris cristata, 34
Iris Family, 34
Ironweed, New York, 110
Ironweed, Tall, 110
Isopyrum biternatum, 50

Jack-in-the-pulpit, 24
Jacob's Ladder, 84
Jeffersonia diphylla, 52
Jewelweed, Pale, 68
Jewelweed, Spotted, 68
Joe Pye-weed, 110
Joe Pye-weed, Spotted, 110
Joe Pye-weed, Sweet, 110
Jug, Little Brown, 44
Juniperus virginiana, 11
Jussiaea decurrens, 74

Karst topography, 4
Knotweed, 42
Kral, Robert, 104
Kuchler, A.W., 5

Labiaceae, 90
Ladies'-tresses, Slender, 38
Lady's-slipper, Large Yellow, 38
Lady's-slipper, Showy, 38
Lamiaceae, 90, 92
Larkspur, Glade, 48
Larkspur, Spring, 48
Leather-flower, 50
Leavenworthia stylosa, 18, 56
Leavenworthia uniflora, 56
Leaves, 14, 15
Lebanon limestone, 7
Lesquerella densipila, 54
Lesquerella lescurii, 54
Lesquerella lyrata, 54
Lesquerella perforata, 54
Lesquerella stonensis, 54
Liatris aspera, 112
Liatris squarrosa, 112
Liliaceae, 26, 28, 30
Lily Family, 26, 28, 30
Linnaeus, Carolus, 18
Liparis lilifolia, 36
Lithospermum canescens, 88
Lizard's-Tail, 42
Lizard's-Tail Family, 42
Lobelia appendiculata, 98
Lobelia cardinalis, 98
Lobelia gattingeri, 98
Lobelia, Gattinger's, 98
Lobelia, Pale Spiked, 98
Lobelia spicata, 78
Logania Family, 78
Loganiaceae, 78
Ludwigia decurrens, 74
Ludwigia peploides, 74
Lupinus, 60

Madder Family, 96
Mallow Family, 68
Malvaceae, 68
Manfreda virginica, 32
Maple, Red, 12
May-apple, 52
May Prairie, 12
Melilotus alba, 78
Melilotus officinalis, 78
Mertensia virginica, 88
Meyer, William, 52
Microclimates, 7
Milk-vetch, Tennessee, 62
Milkweed, Antelope-horn, 82
Milkweed Family, 82
Milkweed, Green, 82
Milkweed, White, 82
Milkweed, Whorled, 82
Milkwort Family, 66
Milkwort, Field, 66
Milkwort, Rosy, 66
Mimosa, Prairie, 60
Mint Family, 90, 92
Mixed Mesophytic Forest Region, 5
Moccasin Flower, Pink, 38
Monarda didyma, 90
Monarda fistulosa, 90
Monocots, 16
Mustard Family, 54, 56
Mustard, Nashville, 56
Mycorrhizae, 6
Myrtle, 80

Nostoc commune, 1
Nothoscordum bivalve, 28

Oaks, 5, 11
Oak-Hickory Region, 5, 11
Oenothera biennis, 74
Oenothera missouriensis, 74
Oenothera speciosa, 74
Oenothera triloba, 74
Onagraceae, 74
Onion, Nodding Wild, 26
Onion, Wild, 26
Onosmodium molle, 88
Opuntia compressa, 72
Opuntia humifusa, 72
Opuntia rafinesquii, 72
Orchidaceae, 36, 38
Orchid, Crested Yellow, 36
Orchid Family, 6, 36
Orchid, Grass-pink, 36
Orchid, Yellow-fringed, 36
Ordovician period, 7
Orobanchaceae, 96
Orpine Family, 58
Oxalidaceae, 64
Oxalis priceae, 64
Oxalis stricta, 64
Oxalis violacea, 64

Paint-brush, Indian, 94
Panax quinquefolius, 76
Panax trifolium, 76
Pansy, Field, 70
Papaveraceae, 52
Papyrus Plant, 22
Parsley Family, 76
Parthenium integrifolium, 100
Passifloraceae, 72
Passiflora incarnata, 72
Passiflora lutea, 72
Passion Flower, 72
Passion Flower Family, 72
Passion Flower, Yellow, 72
Pea Family, 60, 62
Pediomelum esculenta, 60
Pediomelum psoraloides, 60
Pediomelum subacaulis, 60
Pelargonium, 64
Penstemon hirsutus, 94
Penstemon tenuiflorus, 94
Periwinkle, 80
Persoon, C. H., 52
Petalostemum gattingeri, 62
Peterson, Lee, 56
Phacelia bipinnatifida, 86
Phacelia dubia, 86
Phacelia, Fringed, 86
Phacelia, Purple, 86
Phacelia purshii, 86
Phacelia ranunculacea, 86
Phlox bifida, 84
Phlox, Blue, 84
Phlox, Creeping, 84
Phlox divaricata, 84
Phlox, Downy, 86
Phlox Family, 84, 86
Phlox, Glade, 84
Phlox pilosa, 86
Phlox stellaria, 84
Phlox stolonifera, 84
Physostegia virginiana, 92
Phytolacca americana, 42
Phytolaccaceae, 42
Pink Family, 46
Pink-root, 78
Platanthera ciliaris, 36
Platanthera cristata, 36
Pleurisy Root, 82
Pleurochaete squarrosa, 1
Poaceae, 20
Podophyllum peltatum, 52
Pokeweed, 42
Pokeweed Family, 42
Polemoniaceae, 84
Polemonium reptans, 84
Polygala cruciata, 66
Polygala sanguinea, 66
Polygalaceae, 66

Polygonaceae, 42
Polygonatum biflorum, 26
Polygonatum pubescens, 26
Polygonum lapathifolium, 42
Poppy Family, 52
Portulacaceae, 44
Potentilla canadensis, 58
Potentilla recta, 58
Poverty Grass, 20
Prairies, 11-12
Prairie-clover, Gattinger's, 62
Prairie-dock, Cut-leaved, 108
Prairie-tea, 66
Prickly-pear, 72
Primrose Family, 76
Primrose-willow, 74
Primulaceae, 76
Privet, Glade, 11
Psoralea subacaulis, 60
Puccoon, Hoary, 88
Purple-tassels, 62
Purslane Family, 44
Pussytoes, Field, 100
Pussytoes, Plantain-leaved, 100
Pussytoes, Solitary, 100

Quarterman, Elsie, 9, 10, 11
Queen Anne's Lace, 1
Quercus species, 5, 11
Quinine, Wild, 100

Radnor Lake, 3
Ragweed, 106
Ragwort, Golden, 108
Ragwort, Small's, 108
Ranunculaceae, 46-50
Ranunculus abortivus, 46
Ranunculus acris, 46
Ranunculus recurvatus, 48
Ratibida pinnata, 106
Rattlesnake-master, 32
Rattle-vetch, 62
Red-cedar, Eastern, 7, 11
Red Clover, 1
Rhus aromatica, 11
Rollins, Reed, 54
Rosaceae, 58
Rose Family, 58
Rose-mallow, 68
Rose-mallow, Swamp, 68
Rose-pink, 78
Rosinweed, 108
Rosinweed, Whorled, 108
Rubiaceae, 96
Rudbeckia hirta, 104
Rudbeckia laciniata, 106
Rudbeckia triloba, 104
Rue-anemone, 50
Rue-anemone, False, 50
Ruellia caroliniensis, 96
Ruellia, Hairy, 96
Ruellia, humilis, 96
Ruellia, Smooth, 96

Sabatia angularis, 78
Sabatia brachiata, 78
St. John's-wort, Common, 11, 68
St. John's-wort Family, 68
St. John's-wort, Shrubby, 11
Sandwort, Glade, 46
Satureja glabella, 90
Satureja vulgaris, 90
Saururaceae, 42
Saururus cernuus, 42
Savory, Glade, 90
Saxifrage, Early, 58
Saxifrage Family, 58
Saxifraga virginiensis, 58
Saxifragaceae, 58
Schizachyrium scoparium, 20
Schrankia microphylla, 60
Scientific names, 16-18
Scorpion-weed, 86
Scorpion-weed, Glade, 86
Scrophulariaceae, 94-96
Scutellaria parvula, 92
Scutellaria serrata, 92
Sedge Family, 22
Sedge, Frank's, 22
Sedum pulchellum, 58
Sedum ternatum, 58
Senecio aureus, 108
Senecio glabellus, 108
Senecio smallii, 108
Shawnee Hills, 2, 4
Shooting-star, 76
Silene rotundifolia, 46
Silene virginica, 46
Silphium pinnatifidum, 108
Silphium terebinthinaceum, 108
Silphium trifoliatum, 108
Sisyrinchium albidum, 34
Sisyrinchium angustifolium, 34
Skullcap, Showy, 92
Skullcap, Small, 92
Small, John K., 108
Smartweed, Dock-leaved, 42
Smilacina racemosa, 26
Snakeroot, Sampson's, 60
Snapdragon Family, 94
Sneezeweed, Autumn, 102
Sneezeweed, Purple-headed, 102
Soil, 4-5
Solidago altissima, 106
Solidago gattingeri, 106
Solidago shortii, 106
Solomon's Plume, 26
Solomon's-Seal, 26
Solomon's-Seal, False, 26
Spanish-bayonet, 28
Spanish-needles, 104
Spider-lily, 32
Spider-milkweed, 82
Spiderwort Family, 24
Spiderwort, Ohio, 24
Spiderwort, Virginia, 24
Spiderwort, Zigzag, 24
Spigelia marilandica, 78
Spike-rush, 22
Spiranthes cernua, 38
Spiranthes gracilis, 38
Sporobolus vaginiflorus, 20
Spring-beauty, 44
Spring-beauty, Carolina, 44
Spurge Family, 66
Spurge, Flowering, 66
Squaw Root, 96
Squirrel-corn, 52
Star-grass, 32
Stellaria, 84
Stonecrop, Lime, 58
Stonecrop, Woodland, 58
Sumac, Aromatic, 11
Sunflower, Common, 102
Sunflower, Thin-leaved, 102
Sunflower, Woodland, 102
Susan, Black-eyed, 104
Swamp-lily, 42
Sweet-clover, White, 78
Sweet-clover, Yellow, 78
Swertia caroliniensis, 78
Symphoricarpos orbiculatus, 11

Talinum calcaricum, 44
Talinum teretifolium, 44
Teasel Family, 98
Teasel, Fuller's, 98
Teasel, Wild, 98
Temperate deciduous forest, 5
Tephrosia virginiana, 60
Thalictrum thalictroides, 50
Thistle, Bull, 110
Thistle, Musk, 110
Thistle, Nodding, 110
Thrift, 84
Tiarella cordifolia, 58
Tickseed-sunflower, 104
Toadshade, 30
Toothwort, Cut-leaved, 56
Touch-me-not Family, 68
Tradescantia ohiensis, 24
Tradescantia subaspera, 24
Tradescantia virginiana, 24
Trifolium pratense, 1
Trillium cuneatum, 30
Trillium grandiflorum, 30
Trillium, Large-flowered, 30
Trillium, Prairie, 30
Trillium recurvatum, 30
Trillium sessile, 30
Trillium, Sessile, 30

Trout-lily, 28
Trout-lily, White, 28
Turnip, Prairie, 60
Twayblade, Lily-leaved, 36
Twin-leaf, 52
Typha angustifolia, 20
Typha latifolia, 20
Typhaceae, 20

Ulmus alata, 11

Vegetation, 5
Verbena canadensis, 90
Verbena Family, 90
Verbena, Narrow-leaved, 90
Verbena, Rose, 90
Verbena simplex, 90
Verbenaceae, 90
Vernonia altissima, 110
Vernonia noveboracensis, 110
Vinca minor, 80
Viola canadensis, 70
Viola egglestoni, 70
Viola papilionacea, 70
Viola pedata, 72
Viola rafinesquii, 70
Viola striata, 70
Violaceae, 70-72
Violet, Birdfoot, 72
Violet, Canada, 70
Violet, Common Blue, 70
Violet, Confederate, 70
Violet, Dog-tooth, 28
Violet, Eggleston's, 70
Violet Family, 70-72
Violet, White, 70

Ware, Stewart, 44
Waterleaf Family, 86
Water-primrose, 74
Western Mesophytic
  Forest Region, 5, 30
Wild Baby's Breath, 46
Wild Ginger, 44
Witch's Butter, 1, 11
Wood-mint, Downy, 92
Wood-mint, Hairy, 92
Wood-sorrel Family, 64
Wood-sorrel, Price's, 64
Wood-sorrel, Violet, 64
Wood-sorrel, Yellow, 64

Xanthocephalum
  dranunculoides, 104

Yarrow, 100
Yellow Nut-grass, 22
Yucca, 28
Yucca filamentosa, 28

Zonation, Plant, 9, 11